Data Dating

First published in the UK in 2021 by
Intellect, The Mill, Parnall Road, Fishponds, Bristol, BS16 3JG, UK

First published in the USA in 2021 by
Intellect, The University of Chicago Press, 1427 E. 60th Street,
Chicago, IL 60637, USA

A catalogue record for this book is available from
the British Library.

Copy editor: MPS, Limited
Cover designer: Aleksandra Szumlas
Cover image: Ania Malinowska and Valentina Peri
Production manager: Laura Christopher
Typesetter: MPS, Limited
Printed and bound by CPI Group (UK) Ltd, Croydon, CR0 4YY

Hardback ISBN 978-1-78938-495-6
Paperback ISBN 978-1-78938-952-4
ePDF ISBN 978-1-78938-496-3
ePUB ISBN 978-1-78938-497-0

To find out about all our publications, please visit our website.
There you can subscribe to our e-newsletter, browse or
download our current catalogue and buy any titles that are in
print.

www.intellectbooks.com

This is a peer-reviewed publication.

Data Dating

Love, Technology, Desire

Edited by
ANIA MALINOWSKA AND VALENTINA PERI

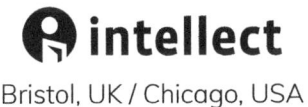

Bristol, UK / Chicago, USA

Contents

Figures

Acknowledgements

We owe a sincere gratitude to many people, institutions and experiences that inspired and supported the book's gestation. First, we would like to thank Galerie Charlot and its founder Valérie Hasson-Benillouche for hosting the *Data Dating* exhibition in Paris and Tel Aviv and for supporting our efforts to extend the exhibition's ideas beyond the gallery space. A big thank you to the Polish-US Fulbright Commission for sponsoring part of the book-related research and to the New School in New York for hosting Ania's research project that significantly influenced the book's development. We would like to thank all our contributors – artists and authors – for their trust, commitment and patience.

From Valentina Peri, a deep expression of gratitude to ZKM in Karlsruhe and curator Lívia Nolasco-Rózsás for inviting her to present the ideas of *Data Dating* in the framework of the exhibition project *Open Codes*; to Watermans in London and its head of new media art Klio Krajewska for hosting the *Data Dating* exhibition; to professor Manuela de Barros at University of Paris VIII for organizing the lecture *Data Dating. Love in the Age of Internet* as part of the Sciences & Fictions cycle and to her family, friends and collaborators for their affection, encouragement and support (you know who you are). A special thanks to Ania Malinowska for her unrelenting optimism and perseverance throughout the process of book production. And from Ania to Valentina Peri for her impressive professionalism, artistic insight and invaluable support and to anyone who has offered assistance and aid to the book at private and institutional levels.

FIGURE 0.1 (pages x–xi): Tom Galle, John Yuyi and Moises Sanabria, *Tinder VR*, 2016. Acrylic print and video (documentation of the performance in NYC). Courtesy of the artists.

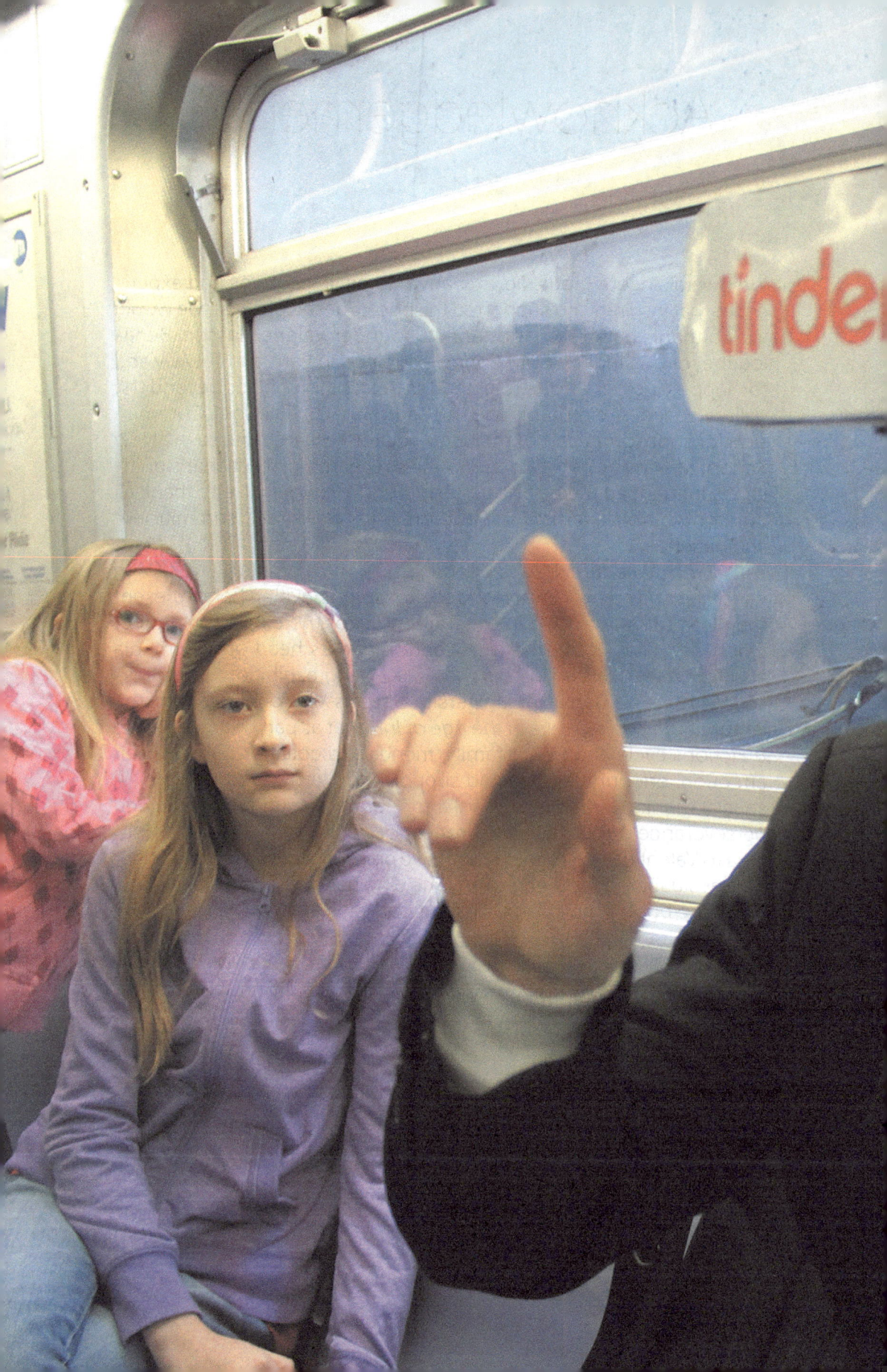

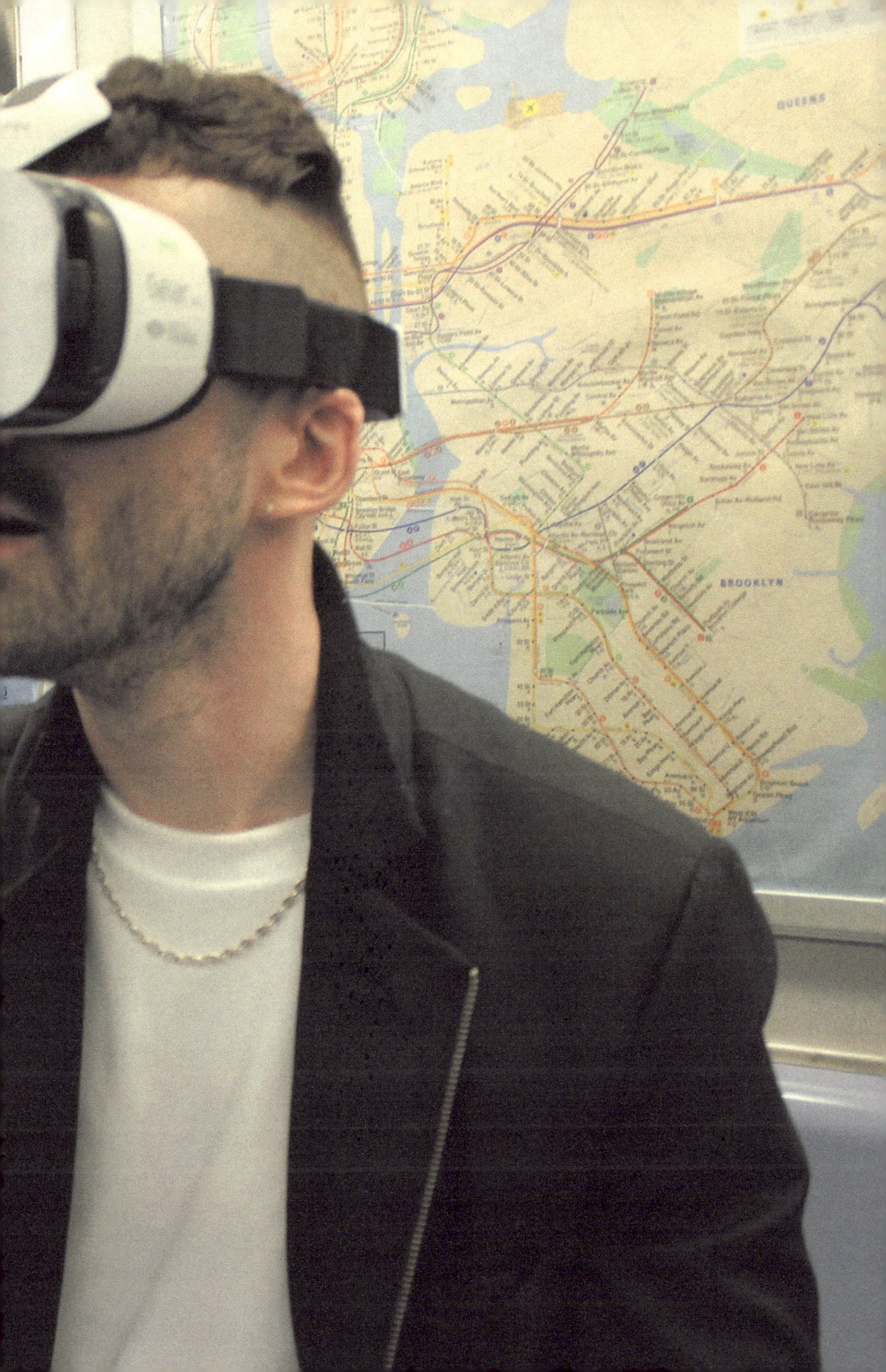

Introduction: Dating (the) Data and Other Intimacies

ANIA MALINOWSKA AND VALENTINA PERI

What does it mean to love in the age of the internet? How are digital interfaces reshaping human interactions? What implications do new technologies carry for the future of our romantic relationships? How does mediation affect our sexual conduct? Or to be more specific, how do screens, gadgets, add-ons, platforms, wearables and other high-tech media artefacts – including technological subjects, like (ro)bots – determine emotional and intimate behaviours that are clearly being remodelled to the demand of new communication formats? Are new digital technologies shifting the old paradigms of love and erotic expression? And with respect to that, can we talk about a change in the ways we practice love or rather we should speak about reformulations of the age-old codes of loving under the new media regimes?

Those questions have been in the landscape since the advent of online media. They became hot-button in March 2020, when a global pandemic placed millions of people under the coercion of a total lockdown, enforcing a transfer of most of our activities to the virtual plane. From online working to online teaching to online voting, humans all over the planet manoeuvred the discontents of social distancing, trying to live the no-contact reality as the new normal. Inefficient for all the ways of living, those efforts turned out particularly futile for intimate interactions: hookups and dating, inspiring both frustration and failed inventiveness. The follow-up debates about the future of romance inquired after the role of technology for human relations in circumstances when the non-contact status is not an alternative but a default. At the same time – as if in response to those discussions – Shinoda and Makino Lab at the University of Tokyo announced the release of the first haptoclones – touchable holograms that enable human interactions at long distances without any equipment. The holograms are likely to revolutionize online communication by allowing touching through haptics across the continents. They also present themselves as solution for social distancing or physical separation – pressured or self-imposed. But will they preserve *the essence of contact* that makes all forms of emotional and intimate bonding a truly magical experience?

DATING

Modern cultural theory censures high technologies for diluting and corrupting 'the event of a human encounter'. The culprits are social platforms and the computational control of a meet-up. Badiou (2012, 2017), who writes extensively about the collapse of loving under the protocols of online dating, claims that *zero-risk policy*, exerted by socially and psychologically biased profiling for an impeccably accurate match, has flattened the romantic pursuit. What sabotages it even further is the culture of sameness bred on 'app-template individualism' and 'marketable privacy'. Bauman (2003) speaks of the destructive unification of romance. As the mystery and alterity have disappeared from dating, love has yielded to the pre-eminence of availability and self-presentation. In *The Agony of Eros* (2017), Byung-Chul Han explains that contemporary technologies transform togetherness – understood as the desire to form a 'we' – into a pleasant symbiosis of self-invested display (a form of 'reciprocal narcissism'). Media environments, Han explains, encourage narcissistic preoccupation with self to the point of killing self's interest in/for the Other. In the media and tech environment, romance is believed to be primarily a scene for achievement rather than exchange; in that scene, the *self* defines itself for another by means of self-presentation (that is by means of what is possible for the self in terms of fashioning it-self for display).

Some media theorists counterbalance that view with claims about the possibility of 'navigat[ing] a world saturated with opportunities for social connections [...] without losing sight of one's self and of losing the sight of others' (Papacharissi 2018: n.pag.). There is a conviction that 'technology can help us reimagine and reinvent how we understand love and life' (Papacharissi 2018: n.pag.). As a major site of practicing togetherness, romance needs reinvention, especially in how it dominates the expression of loving and its social legitimacy. This is the more urgent as all the elements of romance's excitement and drama are being transferred to technological environments with the hope for better efficiency (see Berlant 2012). Technology unveils the cultural flaws of romantic love. It also unveils the futility of hopes bestowed on the devices and gadgets by means of which we seek the magic of love. As Zizi Papacharissi observes,

> [t]ools cannot store bonds that are lost, cannot invent bonds that do not already exist, nor can they restore bonds that are frail. Tinder does not deliver bonds. Match.com does not construct relationships. Facebook does not cure loneliness. Instagram does not make one authentic. We must learn to live with the frailty and insecurity of human existence first, in order to figure out how technology can help us. Love is unpredictable, as it should be, otherwise, where is the charm? Technology cannot predict love, algorithms cannot render it out of databased profiles, and platforms cannot fix love that is broken. *All love stories have a beginning and an end, and technology does not grant them the kiss of eternity.* (2018: n.pag.)

That we cannot know. What we do know is that one in five couples today meet through a dating website. Over 240 million people worldwide use online dating – a number likely to increase in the next decade. That means a potential for profit and an extensive collection of users' data. Dating and hookup applications make the most profitable business in the future of the internet – next to the gaming industry. As of today, romance platforms are ranking third among paid content sites online, outpacing pornography. There is a likelihood of the change in their business model towards recreational romance – an endeavour reflected in the business' preoccupation with bots and artificial assistants. Companion robots and IOS are one of the most rapidly growing branches of robotics. Three in five men would consider a fling with a technologically generated lover. A major shift in our understanding of love is that 'romantic script' can be recreated algorithmically. The roles that people feel they play in romance and that are available for them to perform are reducible to patterns supported by various technologies – organic and artificial, analogue and digital. As Dominic Pettman explains,

> The temptation is to think of love as an intrinsically organic occasion [...]. But [...] all sorts of 'instruments' are required to fuse love back with their (literal) other half [:] the architectural *mise-en-scène*, clothing, cosmetics, cooking, language, techniques of the body, and so on. Humans are saturated by technics, and love is essentially a matter of communication media. [...] it needs technology to do so; whether this is poetry, texting, or Tantra. (2018: 13)

What is, however, peculiar about our times is that they offer a clash of human technicity with the technicity of smart machines. It shows in a constant friction of temporalities (social time vs. digital time), intelligences (computational artificial intelligence vs. human intelligence), environments (online vs. offline) and modalities of presence (logged in vs. unlogged). Those polarities generate frustrations that we try to overcome by emulating the 'human way' onto the machines. Expecting a robot to love like a human is, however, feasible as long as the machine does not hack the coding. Once it does, we cannot be certain of the nature of their sentience, which may not align with our own. Even if much of how we love is founded on 'codes', not all love processes can be reproduced with algorithms inside artificial entities that interact with the world. Once inside a machine, those algorithms may develop beyond the expected effect, producing new forms of intimacy. Perhaps when Deep Love replaces Deep Learning, technological minds become Artificial Loves (and not only Artificial Intelligences). Our task is to embrace this possibility by fathoming the nature and goals beyond sensationalism and anthropomorphic projections.

DATA

This book adopts two critical perspectives: artistic and academic. It gathers new media art works from the *Data Dating* exhibition and theoretical essays from top experts in the field of new media and cultural studies. Preoccupied with the questions of love and technology, it takes those questions away from the polarities of techno-scepticism and techno-enthusiasm and revisits ways in which society is responding to the challenges of technology through connections between emotion, desire, culture, artificial intelligences, digital systems and economy. Organized in ten chapters themed after the economical, psychological, sociological and technological aspects of practicing romance, the book reflects on the structure of feeling(s) and libido environments of contemporary technocracies. The focus is on the problems of new material planes that have emerged from the abstraction of networked communication and virtual landscapes. These are the landscapes where fantasy, fetish and emotions are being shaped by platform capitalism, datafication and new commodity cultures and where self-promotion for coupling nests on the possibilities that come with new media self-mediation formats. Central to the analysis is the carbon-silicon dynamics of love's contemporary DNA and libidinal *techne* (see Pettman 2006, 2020) as practiced in the environment where screens, interfaces, algorithms, data protocols and non-organic objects of affection delineate, organize and program the trajectories of emotional encounter, limerence and erotic pleasure.

The perspectives followed in the book do not limit the understanding of love to romance or adult pleasures. Love is treated here as the entirety of an emotional endeavour by means of which we connect intimately across needs, feelings and imageries. Those connections require reaching out to another as much reaching out to the *self*. How those two are reconciled and compromised in a digitally organized life is a concern of both the artistic exploration covered in the book as well as the critical analysis that extends it. Both of the perspectives critically engage with broader debates on the condition of loving today. However, their themes and areas of critique go beyond the high modern epistemologies of pre-digital and post-digital periodization. Instead, they trace the old social, cultural and psychological patterns of love in the high-tech environments of living to see what happens to those patterns and how they work in the new material reality of digitalism. Most focus is dedicated to the datafication of intimacy, strategies of self-presentation, the kinesis of mediated encounters, digital voyeurism (looking with the media), technological transcriptions of feelings and codification of emotions, virtualizing love discourses, online archives of intimacy and strategies for emotional surveillance. There is also concentration on the actual devices, services and technological objects we use to connect intimately or emotionally and whose meta-forms and meta-functions are being mirrored in the installations, art experiments, high-tech inspired conceptual presentations and other new media artwork that organizes the concentration's thematic and methodological choices.

The book's themes and methods were first rehearsed in the *Data Dating* exhibition – a new media art group show curated by Valentina Peri, exploring new directions of modern romance and sexual identities that this book takes up for further scrutiny (see also datadating.online). The exhibition has been presented at Galerie Charlot in Paris (2018) and Tel Aviv (2019) and at Watermans in London (2020) and continues showing with other contemporary art venues internationally. The featured artists – !Mediengruppe Bitnik, Adam Basanta, Olga Fedorova, Zach Gage, Tom Galle, Thomas Israël, Moises Sanabria, Antoine Schmitt, Jeroen van Loon, Addie Wagenknecht & Pablo Garcia, John Yuyi and Lancel and Maat[1] – offer a number of emblematic insights into the role of new technologies for human emotional engagements, reminding us of the importance of empirical criticism that comes from critically aware practical (menial, physical, spatial, etc.) interactions with the objects of critique – something that the arts do best and something that meta-research invariably misses.

According to some media coverage[2], the exhibition oozes exceptional timeliness. Especially in how it resolves the generational bias over the ways we practice romance these days– a feature attended to by a blogger and digital native Salvia Moon (2020). She wrote, '[the] state-of-the-art, hyperdigital exhibit [...] could easily be branded as a "millennial thing". But perhaps it's a sign of the times; as the internet's sheer penetrative power comes to feel like a fundamental feature of the human condition, we're no longer able to classify digital romance as a discrete generational phenomenon'. French *Libération* (2018) highlighted the experimental level of the artworks included in the show to imply on our own ever 'trial and error' interactions with technology: 'Valentina Peri conceived the exhibition as a form of exploration of emerging practices and current do-it-yourself projects which even the new generations do not always master the rules. [This is] a learning process, [...] which makes the artists' work all the more interesting: their creations both anticipate and question future sentimental and/ or sexual practices'.

OTHER INTIMACIES

The vehicle of the book is the intimate dynamics between the artworks, the artists, the authors and the issues that they all expose and explain. Its major drive, however, is almost libidinal and perverse variety in approaching the same problem and the desire to showing two (or more) sides of one coin even if only for the sake of flipping the perspective.

It starts with an overview of the social codes of romance reflected in automated systems (e.g. platforms). Lauren Rosewarne explains it with an outline of the technological revolution of romance in the macrocosm of online dating to show how emotions and personalities tether to platforms for the formation and consumption of new affective styles and 'limerent' trends. What makes sexting, tuning, cushioning and other mediated hookup baits so appealing is the easy immediacy of affectionate interest they invariably grant. There is nothing more attractive than the instant 'I love you'. Antoine Schmitt's website installation *Deep Love* illustrates this with a visual metaphor of affective unconditionality materialized in a programmed chatbot that is always there for you with a confession.

Chapter 2 continues with a debate on data and their infrastructures as inspired by Zach Gage's *Glaciers*. This conceptual electronic sculpture, digging into the idea of autocomplete: a browser feature offering fixed phrasing for online searches, invites thoughts about the

instability and structured-ness of information, especially their influences on our choices and the contemporary modalities of thinking.
Lee MacKinnon's follow-up examination of dark data links this with
the liquid nature of human intimacy and the exploitation of love under
the market-oriented transfer of desire across dating services. Determined by the selection and flow of data, the game of love yields to the
demands of efficacy. Both information formats and online time-space
environments of online interactions comply with the strategies for
maximizing effects – something that has always been part of romantic pursuits and something that Ania Malinowska reviews in Chapter 3
when talking about human–machine temporalities. Drawing on the
concept of *digitally mediated togetherness*, both Malinowska's feature
on *fast love* and Adam Basanta's kinetic installation *A Truly Magical
Moment* account for time and space travels granted by the virtual
environment of communication media. The influence of those environments on romance shows in the eagerness with which we choose
virtuality and mediation to encompass a number of social, physical
and technological dimensions and enjoy immediacy – across time
zones and busy schedules – that negotiates our ability to manoeuvre
technological spaces.

A different aspect of our co-existence with technological systems
is discussed in Andrew McStay and Gilad Rosener's analysis of empathic
machines in Chapter 4. By asking what it means to feel with a robot –
which they explore on the basis of children's interactions with emotive
toys, the authors guide us through the working of empathic technologies outside to the romantic contexts, shedding some light on how
those early emotional interactions with emotionally responsive automata may condition us for the future. This is particularly precious for
understanding adult interactions with robotic partners and companions as depicted in *Ashley Madison at Work* by !Mediengruppe Bitnik –
a video installation that playfully comments upon the scandal over a
hookup platform manned by bots. Most of the time, we are not even
aware that those we interact with are not humans – like the users of
Ashley Madison services who, looking for people, were duped by the
robots to the point of taking them for real human beings. It is either
we lose our sense for a 'tangible presence' or the tangible presence
of another is no longer of importance. What counts, perhaps, is the
access to interactive fantasy and the illusion of visibility that platforms
and devices create for us. As noted in Chapter 5, desire today is all
about self-love and the mediation of image. In other words, our main
interest is with the presentability of the self to the 'attention-obliging'
virtual void. Either for a hookup (on Tinder) or for polemic (on Twitter),
we throw ourselves to the global network to widely and safely (under

the protection of physical absence) test the potential of our social presence. *Kill Your Darlings* by Jeroen Van Loon, an installation inspired by Twitter and social media youth cultures – examines this in relation to platform violence in young adults. As a major space for interaction, that is, space where young adults form and conduct relationships (be it love or friendship), platforms are a new environment for Bildungsroman – narrating the self. In this space, tweets, shares, texts, posts work as character tools with which the platform participants mould their life persona. Derek C. Murray examines it broadly in a cross-cultural context where platforms serve also for mediating and modernizing racial and ethnic differences and negotiating non-western cultural styles.

Chapter 6 discusses the balance of display and visibility through the changing notion of the obscene – once associated with pornography and today with social presence and belonging. Nudes are no longer the core of the voyeuristic desire. Intimacy is not that much about the organs but about the orchestration of private modes of presentation – now more amateur than ever and perhaps so intuitively primal. Being looked at seems to be more of an appeal than looking at. *Peeping Tom (Porn Version)* by Thomas Israël grasps that preference in a video installation broadcasting an obtrusive eye – the organ of penetration in the era when physical contact gains different senses. Lynn Comella traces the change in her analysis of the new obscene that emerged with the forms of mediating pornography. A key to understanding obscenity and its related intimacy today is understating the shift in our perception of proximity and what distance means in the environment where physical immediacy is separated with screens, camera lenses, interfaces, etc. As shown in Chapter 7, mediation affects our reactions to space and alter our behaviours. In mediated interactions, we are either bolder, less tuned up, or bolder and totally overdone. Addie Wagenknecht and Pablo Garcia picture that in *Webcam Venus* – a media-art social experiment with sexcam performers who were asked to pose after a chosen classic painting. Not only does it provide an insight into broadcasted intimacy but also portrays styles of reconciling private lives and sex work. Kyle Machulis explains it in the analysis of digital proxemics that provides an understanding of how people perceive the inhabitants and features of virtual space in romantic and intimate relationships. The follow-up Chapter 8 extends this with David Parisi's study of haptic media based on hug t-shirts and other touch technologies – the phenomenon also presented, quite ironically, in *VR Hug* by Tom Galle and Moises Sanabria. Hug t-shirts represent a tactile technology believed to materialize physical contact in virtual interactions. Parisi's contribution explains the infrastructure of transmitting hugs, the semiotics of encoding digital embraces and the role of networked affective communication in what we may term 'the general crisis of touch'.

Chapter 9 puts in words how touch – so far translated into signals and vibrations to transmit physical presence – is also translated into sound. *EEG KISS* by Lancel & Maat, a science-ridden audio-visual performance, presents technological solutions for sonifying an act of kissing and representing it with music and image. Sound expert Andrew Blanton articulates the working of the experiment, giving us another insight into how technology feels us and how we feel it in response.

The concluding Chapter 10 discusses social surveillance and the importance of control in romantic relationships, elucidating the discontents of wired togetherness. As Nanna Bonde Thylstrup and Kristin Veel explain in their account of time and geostamps, the penchant for monitoring behaviours of our significant others and knowing their moves has always been a vital part of the game of love. But never have the tools of control been more available than they are now. Surveillance features in communication devices accelerate romance: helps to manage its development and determine its trajectories. It also significantly frames our emotional responses, causing distress that sabotages the romantic process. John Yuyi and Tom Galle's *Face Messenger* renders the pain of online control in a close-up of an eye tagged with 'Seen' stamp.

Despite their distinct sectioning, the chapters overlap in terms of problems, examples and methods for revealing different aspects of dating in the media-saturated reality. The uniqueness of their approach is reflected in how they navigate and balance high-tech practices and cultural concerns around love. Crucial for that matter is the engagement with media art, used not as an illustration of critical points but as an indicator of trajectories for theorizing love in relation to technological/new media phenomena and vice versa. Faithful to anthropological and sociological methods, the book extends them with its focus on somatic, performative, affective as well as material and physical aspects of feeling and behaving romantically with/via technologies. The proposed combination of academic (textual) and artistic (conceptual) perspectives unleashes a more holistic and analytically broader meta-commentary in which love and intimacy, observed in a wide spectrum of a critical practice, reveals itself as inventive phenomenon aware of its technics.

References

Badiou, Alain (2012), *In Praise of Love*, London: Serpent's Tail.

Badiou, Alain (2017), 'Foreword: The Reinvention of Love', in H. Byung-Chul, *The Agony of Eros*, Cambridge: The MIT Press, pp. vii–xi.

Bauman, Zygmunt (2003), *Liquid Love: On The Frailty of Human Bonds*, Cambridge: Polity.

Berlant, Lauren (2012), *Love/Desire*, New York: Punctum Books.

Byung-Chul, Han (2017), *The Agony of Eros*, Cambridge: The MIT Press.

Giard, Agnès (2018), 'Voulez-vous 30 seconds d'amour fou?', 7 September, http://sexes.blogs.liberation.fr/2018/07/09/voulez-vous-30-secondes-damourfou/?fbclid=I-wAR0DiKefjun6l6FrQVcmKWnDkZfT-ke6RA6rj_rJEvGuxnhahlwmjxg8kkvo. Accessed 20 February 2020.

Moon, Salvia (2020), 'Art that returns the gaze: *Data Dating* review', 29 January, http://salviamoon.blogspot.com/2020/01/. Accessed 1 February 2020.

Papacharissi, Zizi (ed.) (2018), *A Networked Self and Love*, New York: Routledge.

Pettman, Dominic (2006), *Love and Other Technologies: Retrofitting Eros for the Information Age*, New York: Fordham University Press.

Pettman, Dominic (2018), 'Love materialism: Technologies of feeling in the 'post-material' world (an interview)', in A. Malinowska and M. Gratzke (eds), *The Materiality of Love: Essays of Affection and Cultural Practice*, New York: Routledge, pp. 13–24.

Pettman, Dominic (2020), *Peak Libido: Sex, Ecology, and the Collapse of Desire*, Cambridge: Polity.

Urry, John (2008), 'Speeding up and slowing down', in H. Rosa and W. E. Sheureman (eds), *High Speed Society: Social Acceleration, Power and Modernity*, University Park: Penn State University Press, pp. 179–200.

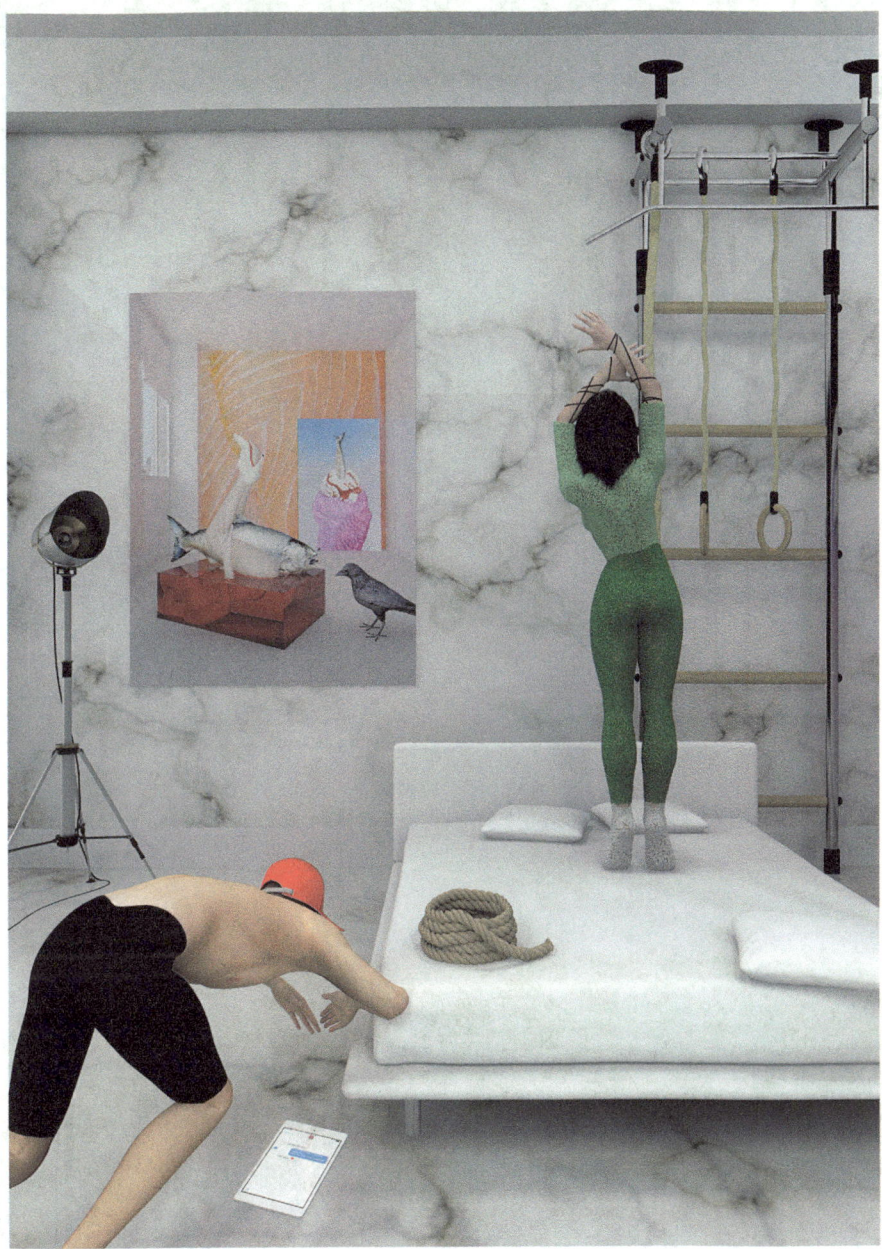

FIGURE 0.2: Olga Fedorova, *The Myth of Female Solidarity*, 2017. Lenticular print and animated GIF. Courtesy of the artist and Galerie Charlot, Paris/Tel Aviv.

FIGURE 0.3 (pages 26-27): Olga Fedorova, *Green Room*, 2017. Print. Courtesy of the artist and Galerie Charlot, Paris/Tel Aviv.

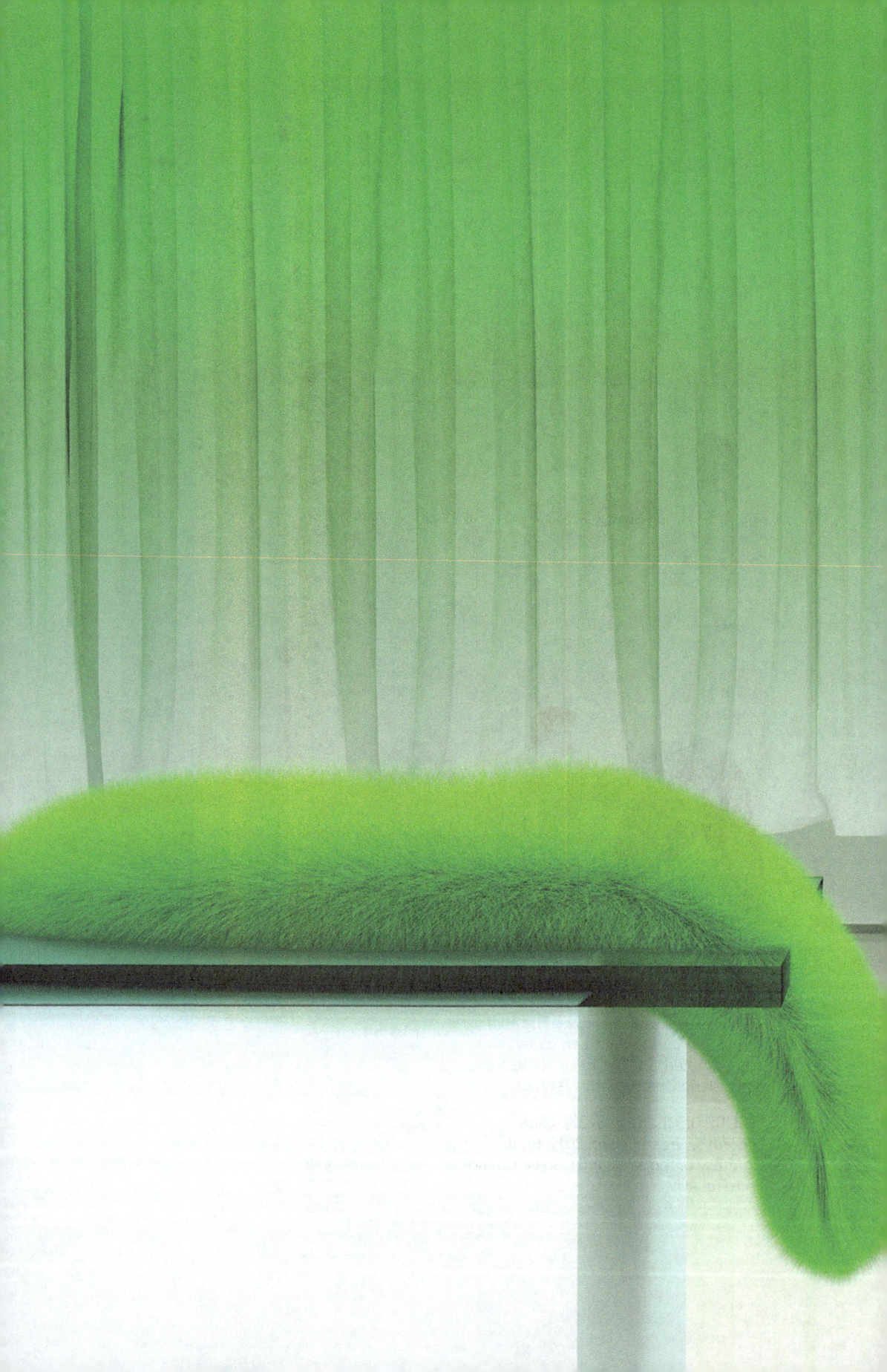

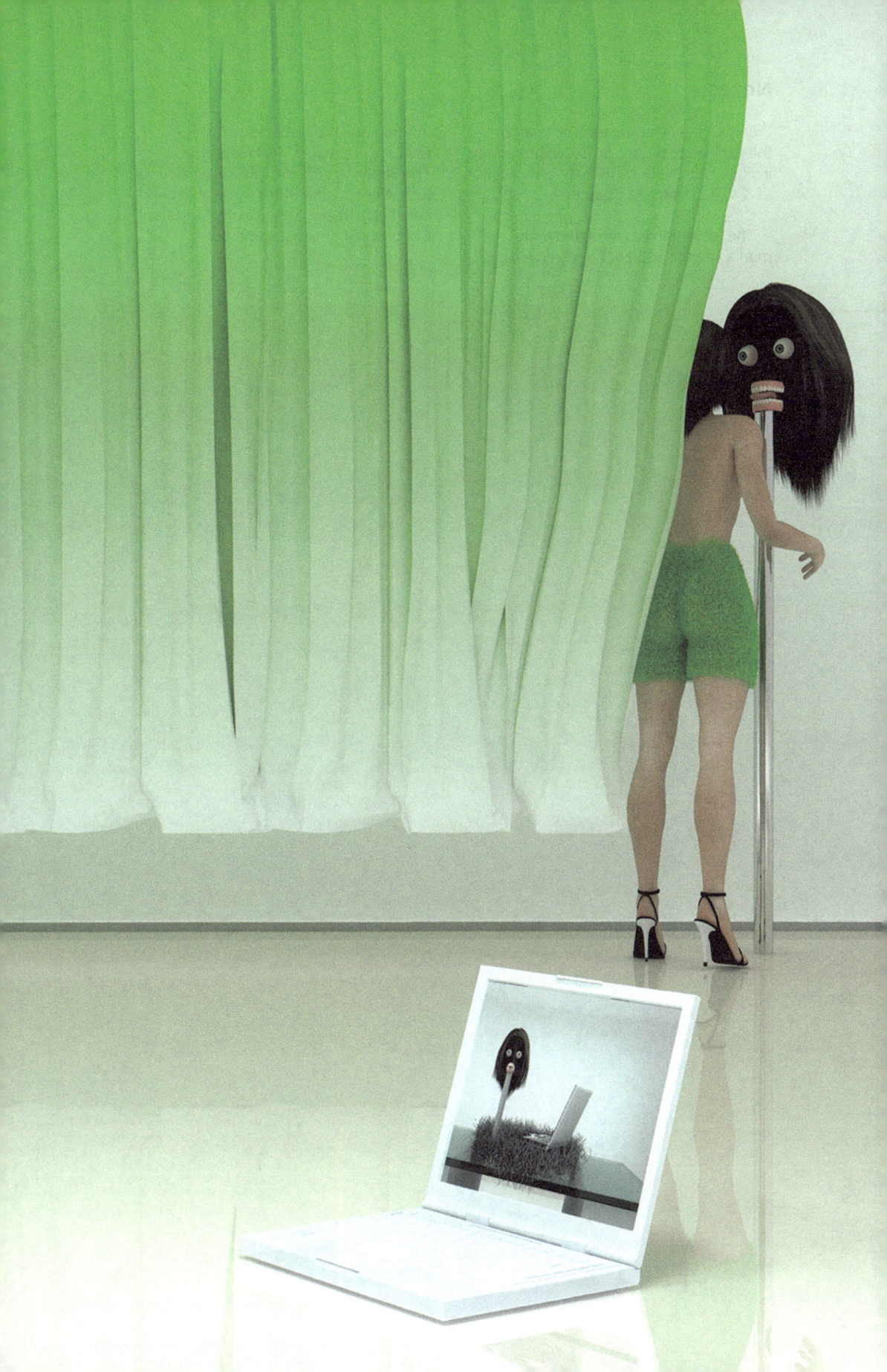

Notes

1. Lancel and Maat's work is not part of the exhibition, but the artists accepted our invitation to the book with their installation *EEG KISS* covered in Chapter 9.

2. The exhibition has attracted large media attention and has been covered internationally by Libération (three features), Télérama, *FAD Magazine*, The Steidz, Art Rabbit, L'ADN, Usbek & Rica, Neon, Contemporary Lynx, Digi-cult, Furtherfield, Arts Hebdo Medias, Le Bonbon, i24 channel and others.

1
Wired Limerence

Network reality has created conditions of an almost unlimited connectivity. We can now access people, locations, services and goods from any place on the planet with connection to wide bandwidth data transmission or any high-speed internet. Not only has this changed our understanding of distance and communication but also wired us to a pool of options far beyond our immediate environments of choice. Romance, which thrives on an easily accessed variety, tests that pool with technologies of togetherness that marry the abundance of possibility with what we seem to desire most: unconditional love. Online dating in its many incarnations – from sites to apps to virtual companion devices – networks millions of people, media forms, discourse protocols and operational algorithms with an aim to satisfy human need for flirtation, closeness and emotional and/or physical intimacy. The tangle of objects, interfaces and people – coded for the experience of loving (being loved and giving love) is exactly what we sign up for in our human-to-human or human-to-system romantic interactions.

Artist Antoine Schmitt metaphorises this experience in his website experiment *Deep Love* (2017). Designed as a chatbot run by the Artificial Mind, *Deep Love* interacts lovingly through a text dialogue. Once you open the site, you see a conversation box that welcomes you with a 'Hello, I love you' greeting. Regardless of what you reply, the Artificial Mind always responds with 'I love you' and continues in the following manner:

Deep Love: Hello, I love you.
User: I love you back.
Deep Love: I thank you, and I love you.
User: Thank you.
Deep Love: I don't understand everything, but I love you.

The Mind behind *Deep Love* represents pure unconditional feeling, reciprocated besides reason and consciousness. It only knows you through words, as words are its only reality. The interaction with the Mind is a reminder that wherever love takes its energy and whatever processes it entails, it can be easily simulated, i.e. reproduced on a computer (a Universal Machine in a Turing's sense) by means of algorithms that are the dynamic material of computers. The depth of the simulation may vary from basic (simple textual interaction), to intermediate (simulations of behaviours, movements, thoughts), to very advanced. A good example of the latter is Schmitt's another project *CliMax*, done in collaboration with performance artist Hortense Gauthier. *CliMax* imitates deep hormonal and cerebral processes of pleasure in animate and inanimate organisms. It renders it by means of a metaphor of a hybrid erotic encounter between an artificially generated hologram organism and a human female.

Whatever the depth of the algorithmic simulation, these loving entities exist in reality: they contain inner love processes and express love to the outside world. Controversial as it may seem, such an approach counterbalances the dark vision of malicious AI and the caricature of human cohabitation with artificial life most often presented it in terms of competition, dominance or the mutual exploitation of organic and non-organic species. *Deep Love* offers an alternative depiction: one that questions the nature of feeling and the nature of human emotional capacity. Is love indeed as natural as we tend to insist? Is technology-facilitated intimacy less authentic? The online love industry showcases deep constructedness of the human idea of romance and companionship. At the same time, it reveals a growing need to connect. Technologies we employ to increase our chances for limerence grant connectivity. They may also, however, trade the opportunity for the experience of unconditional love, even if wired by codes.

Deep Love

ANTOINE SCHMITT

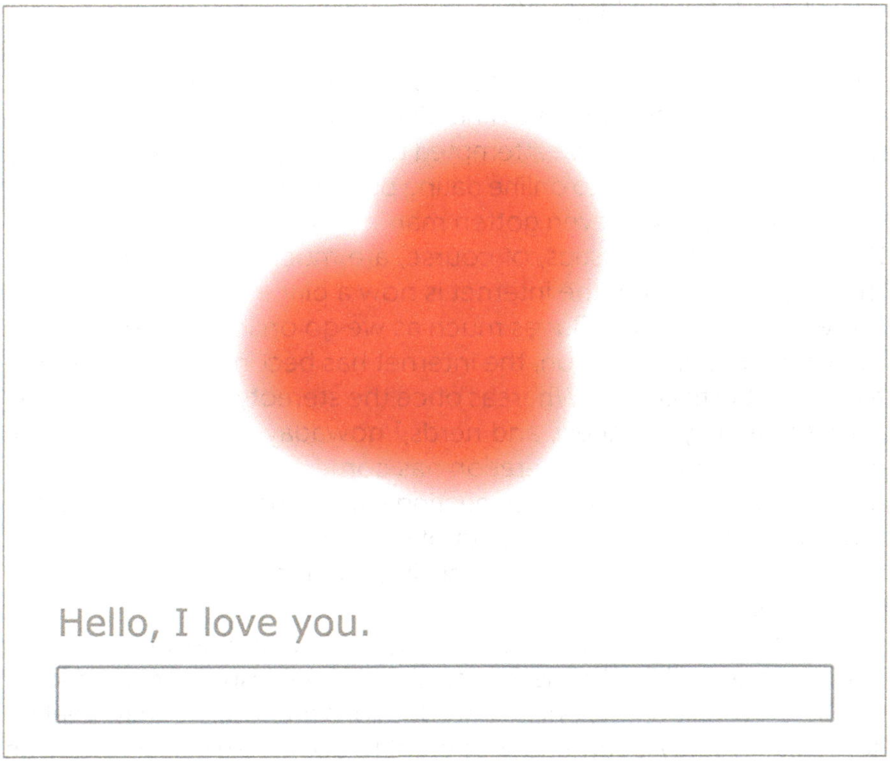

Hello, I love you.

FIGURE 1.1: Antoine Schmitt, *Deep Love*, 2017. Website. Courtesy of the artist and Galerie Charlot, Paris/Tel Aviv.

Technology, Commerce and the Intimacy Revolution

LAUREN ROSEWARNE

From the early 1990s' bulletin boards to today's hookup apps, people have been using the internet to find friendship, love and sex for decades already. Many studies have attempted to put a figure on just how many people have signed up to online dating platforms, dated someone they met there, had sex or even gotten married as a result (Cacioppo et al. 2013; Hall 2014). Statistics, of course, are much less important than the simple reality that the internet is now a place – if not *the* place – turned to for intimacy. But as much as we go online to seek answers to every other life question, the internet has become the default way to connect with others. Whereas once the stereotype was that dating online was only for geeks and nerds,[1] nowadays everyone is online all the time – an entire generation has come of age with this method being the primary means of connecting – and thus the very technology that we use for every other aspect of our life has been well integrated into our intimacies, in turn normalizing and mainstreaming the *who* of online intimacy-seekers.

While it has become increasingly likely that a new romantic or sexual partner will first be met online, there is a widespread perception – despite how ubiquitous online introductions are today – that meeting someone this way is somehow different to doing so elsewhere; that an online connection is dissimilar to meeting through work, at school or church, or via the idealized 'meet cute' so popular in film and television. Culturally, we make assumptions that the people encountered online are less trustworthy, more predisposed to duplicity and perhaps even more dangerous (Rosewarne 2016a; Rosewarne 2016b) – that the partner met online is more likely to harass if not commit outright violence – and we envisage the online meeting as less 'real', less genuine and less romantic than offline introductions. In many ways, this perception *is* the reality: the way we search for, promote ourselves, upgrade and dump people met online does indeed expose some fundamental ways that online-sourced intimacy *is* different. While for digital natives, the dating landscape has always been this way, for older users, the dating world has been revolutionized, with a line sharply dividing the world before and after the internet. A key driver of this difference – and the focus of this chapter – is commerce and the internet as a place for trade just as much as it is for human connectivity, and thus some of the values of the market have resulted in a reconceptualizing of intimacy-seeking.

Thinking of the internet as a quasi-physical place has been common since the inception of the technology. With early popularity of terms like *cyberspace* and *information superhighway*, and commentary referring to the internet as the *wild west* (Rosewarne 2016a), the idea of people being *from* the internet, or *going* online to do or find something, continues to shape our perceptions of the internet as a *where*. This where has been understood in different ways: *wild west*, for example, alludes to ideas of lawlessness and the internet as potentially a kind of *badlands*, and *cyberspace* taps into a futuristic, exciting unknown, if not even a *techno-Utopia*. Of all the different kinds of spaces that the internet provides – spaces to socialize, to politick to seek knowledge – the idea of the internet as a *marketplace* has become central to both our imaginings and also our daily use. For several decades, we've become highly skilled at browsing and buying to such an extent that the internet has completely overhauled retailing and posed a significant challenge to bricks and mortar stores often unable to compete with the ease of shopping in our pajamas in the middle of the night and having our purchases delivered to our front door. Online shopping for goods and services has, in part, trained us for a kind of shopping for intimacy whereby similar skills and a similar ethos are applied. This chapter focuses on online intimacy-seeking being normal and mainstream but with fundamental differences – specifically differences inextricably linked to buying, selling and consumption to intimacy sourced offline. This discussion begins with a brief overview of the dominance of technology in the human connectivity arena.

GOING ONLINE: THE ONLY GAME IN TOWN

In sociologist John Bridges's work on online dating, he observes that dating sites have not only become the accepted way to meet new people, but that 'some really do see it as the only game in town' (Bridges 2012: 43). Bridges's work predates the rise of hookup apps, and yet his ideas have only grown in relevance with the surge in popularity of apps such as Tinder, Grindr, Bumble and Fetlife, each having expanded the concept of online dating to include not only introductions but hookups and playdates geared around certain niche sexual interests. The idea of going online as the default way to meet someone is reiterated in advertising whereby online dating companies actively tout the popularity of their sites, in turn working to normalize their use. eHarmony's advertising, for example, frequently references success stories, with their website boasting, 'We've helped over one million people create successful relationships and understand what it takes to initiate harmony between two people'. Match.com runs similar advertisements claiming that 25,000 people join their site every day. Match.com once aired a commercial alleging that their matches lead to more second dates and marriages than any other. Widespread participation means that online intimacy-seeking today is largely perceived as mainstream rather than something utilized only by a niche and nerdy subculture as it once was (Slater 2013a).

There are several possible explanations for why the internet has become so dominant in the intimacy landscape. First, the internet does not create in us new desires. While the technology certainly helps us do certain things more efficiently – finding love, making friends, having sex, as well as bragging about our relationships and mourning their loss – these are things that we have been doing offline for time immemorial; our methods may have changed, but the desires remain the same. The internet has thus become the default tool used in this space primarily because it helps humans to do what we have always done using a method that makes the process easier, cheaper and more efficient. Other explanations, however, also exist. Smartphones offer us not only permanent connectivity, but have become a central part of our recreational lives whereby searching and scrolling has become a key leisure activity. Extending our searching and scrolling to include the search for an intimate partner has thus been effortlessly integrated into our lives. Coinciding with the rise and dominance of the internet and, notably, online socializing, is the decline in participation in other kinds of offline activities that once commonly introduced us to our partners. As political scientist Robert Putnam observes in his seminal book *Bowling Alone* (2000), we've become a society that's less likely to physically join things, a situation created, in part, because of the internet enabling us to do so many things completely on our own and without leaving our homes. Further, with venues like discos and nightclubs on the decline (Mathieson 2019), and with the concept of single bars seeming dated, the internet has stepped in to provide a comparatively painless means of initiating intimacy. While dating online is low cost compared to many other introduction methods, for many users, it is also perceived as less intimidating, whereby 'the face-to-face judgements that often plague people in real life who exist outside of society's notions of "attractive"' get delayed (Rosewarne 2016b: 85). In light of the changing social scripts pertaining to gender relations as well as the rise of MeToo and heightened concerns about sexual impropriety and sexual harassment (Rosewarne 2018a), initiating a first date online can simply seem like a less confronting option. Social theorist Sophia De Masi also proposes a range of demographic factors that have led to the embrace of the internet in this space including the postponement of marriage, high divorce rates, longevity and the rise in acceptability of homosexuality, each having increased the dating pool and online options being embraced out of sheer necessity (De Masi 2006). Diana Senechal in her book *Republic of Noise* (2011) makes a similar point in the context of widespread loneliness driving the industry – something that perhaps ironically can be connected to our use of the internet itself and the fact that, as noted, it offers us recreation that is largely conducted independently and also *privately*.

In the sections that follow, I focus on ways that the skills honed through online shopping have shaped our intimacy-seeking, beginning with a discussion of 'going to market' in search of love and sex.

SHOPPING FOR LOVE AND SEX

In his book *Love Online*, sociologist Jean-Claude Kaufmann bitingly remarks, 'Welcome to the consumerist illusion which would have us believe that we can choose a man (or a woman) in the same way that we choose a yoghurt in the hypermarket' (Kaufmann 2012: 6). Journalist David Masciotra offers a similar analogy: 'Online dating offers transactional romance, allowing users to browse for a partner as they would browse for a book, refrigerator, or lawnmower' (Masciotra 2015: 175). Amid their cynicism, Kaufmann and Masciotra encapsulate both the perception and the modern reality of online intimacy-seeking being in many ways similar to online shopping.

Online shopping has trained us on navigating infinite choice. Online mega marketplaces such as Amazon, eBay and Alibaba have put the entire world's retail inventory at our fingertips, and out of both convenience and also necessity, we've had to devise ways to traverse this space. We have learnt, for example, the importance of narrowin g our searches by conjuring lists of specific wants. Seeking out companionship with a list of desired qualities – ticking online boxes, for example, as related to diet, religion, pets and zipcode preferences – has become a normal way to narrow our searches. Doing so, however, has been criticized as lacking in romance and serendipity and as bucking notions of destiny (a topic returned to at the end of this chapter).

Despite the widespread cultural belief in the importance of destiny and patience in the quest for intimacy, the existence of the technology – and the ability to be more active about meeting someone – is indicative of self-determination and evidence of individuals becoming consumers in the intimacy market. In signing up to a site or by downloading an app, a person is (a) making a conscious effort to change his or her intimate life and (b) actively outing themselves as someone who is looking to connect as opposed to passively waiting for a serendipitous meet cute. Here, just as we go to the market in fulfilment of other needs – from groceries to theatre tickets to an appointment with a hair stylist – dating platforms are there when we decide to try to meet someone and not just wait for it to spontaneously happen. Ideas around proactivity and agency are well illustrated by an online dater quoted in Dan Slater's book *Love in the Time of Algorithms*, who had seemingly internalized many of these notions: 'My friends think I'm crazy for being on Match [...]. They tell me I don't need it. But quote unquote need isn't really the point, is it? It's about taking control' (2013a: 6).

One key aspect of online intimacy-seeking that links the practice to online shopping is the making of split decisions, often on scant, incomplete or potentially outright deceitful information. Offline we commonly examine, try on and test drive products, or – as related to intimacy – get to know a person in a social setting before determining whether or not sufficient spark exists to initiate a date. Online, however, a choice – be it related to shopping for a product or searching for a partner – is often made on the basis of limited or even inaccurate information.

Split-second impulse buys

Online dating is often criticized as a distinctly shallow practice (Anderson 2016). Because of the catalogue-like display of thousands of photos of fellow intimacy-seekers, site users are searching and scrolling and making decisions about attraction based on a mere glance at a thumbnail. Such a scenario is often criticized as callous and superficial and as resulting in certain types of conventionally attractive people fairing better on these appearance-centric platforms than others. In psychologists Rosanna Guadagno, Bradley Okdie and Sara Kruse's work on dating preferences for example, they identified that 'short men and overweight women were the least likely to get emails through the dating site' (Guadagno et al 2012: 643). Diane Mapes hinted to something similar in her personal development book *How to Date in a Post-Dating World*. Diane Mapes hinted to something similar in her personal development book *How to Date in a Post Dating World*. She observes that:

> [a]ccording to Judy McGuire, who writes the syndicated column 'Dategirl', it's simple. 'Women worry that the guys they meet on the Internet are going to be serial killers. Men only worry that the women are going to be fat'. (In Mapes 2006: 106)

While a thin woman/tall man bias may haunt online dating platforms, such aesthetic preferences are equally pronounced offline with exactly the same gender attributes sought regardless of where a match is solicited. Equally, attraction is integral to *any* connection and therefore while appearance might determine romantic success online, the same situation occurs offline as well: we are a visual culture, preferences around appearance abound and well-meaning adages about not judging a book by its cover have limited relevance in any dating market. While in real life we might find ourselves unexpectedly turned on – or off – by more nebulous factors like a laugh or a gait, the role of attraction has always been fundamental in any kind of dating: in real life, we make the same rapid-fire decisions about attraction and hence the perceived importance of *first impressions*. That said, the internet makes the idea of a first impression more complicated. First, unlike in real life where first impressions are gleaned from meeting a person, interacting with them and then seeing if sparks fly, a first impression online is more likely to be highly orchestrated. In economist Paul Oyer's book *Everything I Needed to Know about Economics I Learned from Online Dating* for example, he discusses orchestration as related to his own profile and his own efforts to tailor his pitch for the dating marketplace:

> [I]f I revealed my [favorite Internet] video and occasional smoke in my profile, that woman would never agree to meet me in the first place. So I make an executive decision. I, like many others, hide these minutiae. (Oyer 2014: 27)

Here, Oyer hints to the importance of *curating* a presentation of self online and the conscious efforts made to manipulate a good first impression which, while having similarities with how people act in real life – i.e. by dressing up or using make-up – doing so online can involve editing and a form of self-branding used to facilitate an electronic introduction in ways that can rarely be accomplished offline.

Branding and self-promotion

With the online dating landscape so crowded, a burden exists to make yourself – *your product* – stand out from the crowd. Oyer discusses this need to self-promote as rendering online intimacy-seeking akin to the other kinds of promotions in the marketplace:

> Sure, there are a lot of differences between someone selling a used bowling ball on eBay and someone signing up for

Match.com, but the basic idea is the same. The bowler needs to think about how to present his bowling ball to get what he wants (money, presumably) just as the Match.com participant needs to present himself to get what he wants (a partner in most cases, casual sex in others). (Oyer 2014: 2)

While all methods employed to find intimacy necessitate some level of self-marketing – even simple grooming is an example of self-marketing rituals routinely undertaken offline to positively present oneself to the world – this situation is amplified online. Simply signing up to a site adds oneself to the electronic catalogue and positions oneself in the hope of being chosen. Self-promotion is amplified online because a photo and a short text pitch are the central way to make an impression and to secure matches and thus has direct impact on success; as Bridges notes, 'the profile is initially the only way to "get one's foot in the door" and have a chance at a potential sale' (Bridges 2012: 66). Choice of photo and creativity with a text pitch thus become crucial in positively positioning your product in the intimacy marketplace.

The online dating participants in sociologists Jo Barraket and Millsom Henry-Waring's study appeared to have *internalized* the necessity to self-promote with one dater going so far as to explicitly see herself as a product ready for consumption and as bearing the burden of carving out a unique profile in a marketplace of similarly packaged goods:

> I mean, you're a product on the shelf, literally on the shelf most of the time, and no-one's ever going to buy you if that's how you present yourself, both your picture and your profile. I mean they're so unimaginative that they talk about walks on the beach and romantic candlelit dinners, all of them. (In Barraket and Henry-Waring 2008: 162)

The idea of putting one's best foot forward can result in the kind of editing that Oyer described, as well as embellishment that might even be construed as lying, alternatively, as self-spin: as psychologists Monica Whitty and Adam Joinson ask in their work on online dating: 'Lies or a different presentation of self?' (Whitty and Joinson 2009: 80). Deceit online is not always engaged in for the purposes of being predatory, but rather, simply to sell oneself in the crowded intimacy marketplace that invariably amplifies the biases and preferences of the offline dating world and where selection or rejection is happening constantly: being online after all introduces us – and sees as being selected or rejected – many more times than ever would occur offline. In order to make a positive first impression, a flattering (and, potentially, outdated) photo might be used and details about height and weight might be tweaked out of an awareness of established appearance biases, all with the intent of just getting a foot in the door. Arguably such orchestrated presentations are so normalized and are so readily assumed anyway that it arguably leads to *more* lying and also *normalized* lying of the *everybody does it* kind; something Oyer explained: 'profile-inflaters [...] have made it seem that because everyone is lying a little, to claim "A Few Extra Pounds" would mean one is actually significantly over-weight' (Oyer 2014: 30).

Of course, sometimes deliberate and distinctly *malevolent* lies are told. Be it through trolling or catfishing, or attempts to fleece via a romance scam (Rosewarne 2012) or to find victims for other crimes, the internet is often looked at as a place disproportionately danger-ous, something fuelled by the ability to remain anonymous for longer and to use some of the editing and profile-manipulation techniques discussed earlier. Doing so, notably with multiple victims at once, is made easier by the technology even if, I would argue, such fears tend to be sensationalized by a techno- and cyberphobic media.

Something hinted to by Whitty and Joinson, and another aspect of online self- marketing, is the capacity to play with identity. While this idea can be interpreted as a variation of the deceit discussed earlier – and can even be done for malevolent purposes as discussed earlier – it can also be construed as another aspect of self-marketing and as something done to broaden one's appeal.

Identity-play

Through our ability to hide behind a screen and remain anonymous for as long as we choose, the internet provides the opportunity to try out different personalities, alternatively, to perhaps amplify traits normally downplayed. In sociologist Aaron Ben-Ze'ev's work on dating online for example, he posits, 'Cyberspace is similar to fictional space in the sense that in both cases the flight into virtual reality is not so much a denial of reality as a form of exploring and playing with it' (2004: 3). In sexual health researchers Danielle Couch, Pranee Liamputtong and Marian Pitts's study on online dating, the authors discuss Cameron who 'held two different online dating profiles on one website to intentionally attract different types of women' (2012: 705). Bridges identified something similar, observing that 'many individuals maintain membership on several different websites, especially given that not all dating sites work the same way in present-ing partners, and not all market the same end result' (2012: 2). In each of these examples, individuals audition different kinds of self-marketing techniques – potentially via different platforms – in an attempt to broaden their appeal. While these different profiles can potentially contain deceitful content, arguably they may be construed as containing different versions of self: an *idealized* self, perhaps, or a self with attributes amplified that normally get downplayed.

While identity-play can be a way to solicit more matches, it can also be an audition of different *sexual* identities without actually committing to anything.[2] Ben-Ze'ev expands on this idea:

> In offline personal relationships, such as marriage, there is less room for mistakes: one or several significant mistakes may wound the spouse in a way that will terminate the relationship or severely harm its quality. (2004: 34)

Meeting someone online for the purposes of a liaison that might be non-vanilla – for example, homosexual or perhaps related to BDSM – can be a way to try a sexual activity without actually committing to an identity centred around that practice. Meeting someone online can create the capacity to contain the encounter, compartmentalize it and perhaps decide to leave it as a one-off sexual experience, as opposed to something put at the centre of life and identity (Rosewarne 2011); alternatively, it can motivate the embrace of a new sexual subculture. The internet can anonymously connect people for these purposes in ways that has always been much more difficult – and risky – to accomplish offline. Going online creates the opportunity to play with identity – to try out that new sexual personality, perhaps – without risking one's offline relationship, reputation or even sense of self.

While, as noted, the internet can connect people for the purposes of homosexual acts or BDSM, more broadly the internet brings together people seeking attributes in a partner that are much more difficult to find offline.

Niche dating

Since the earliest days of the internet – and something that has only exacerbated since – the technology has been used to connect people with similar interests, no matter how fringe. While initially those interests centred on computing (Slater 2013a), the technology very quickly became used to connect people with sexual interests of the kind that are unlikely to be found in someone met by chance offline. Being able to go online and be highly specific about a search can help to locate that intimate needle in a haystack; something that Slater explains well:

We live in an age of customization. Homes, cars, vacations, college degrees, even children – they can all be specialized and designed to suit every need, fad, desire, and whim. Love and relationships have also become customizing things, defined in a boardroom, built to fit, and pushed out by sites like eHarmony and Chemistry, or offered up by OKCupid and Match as one choice among many, as a box that can be checked along with other boxes, such as long-term dating, short-term dating, new friends, casual sex, activity partners. (2013: 115)

Customizing a potential partner positions the online search for intimacy as ever closer to Kaufmann's analogy of obtaining yoghurt or Masciotra's purchase of a lawnmower – and, consequently, makes the exercise seem somewhat prescriptive and unromantic – but can also be considered, simply, as just a way to expedite the process and filter out unsuitable prospect. While searching for someone who shares your sexual kinks is one aspect of this – Fetlife, for example, centres its mission on connecting people for the purposes of fringe sexual activity – as De Masi observes, the internet has also been highly influential in helping people connect on ethnic and religious grounds as well (De Masi 2006).

Mentioned earlier was searching and swiping being a leisure activity. The notion of the market helping to shape pastimes – notably in the realm of intimacy – is by no means a new phenomenon. Social theorists Ida Johnson and Robert Sigler, for example, identified that, 'By the beginning of the twentieth century, dating as recreation was beginning to emerge as an activity not necessarily related to courtship' (Johnson and Sigler 1997: 17). As related to online intimacy-seeking, this idea is well illustrated through swiping through an electronic catalogue of faces now being its own kind of leisure activity. Equally, while meeting someone and going out has long been an aspect of human socializing, the internet has made this all much easier to organize: with the aid of the internet we're also doing it more often and with more people.

Searching and swiping as recreation

Whether online dating companies have caused, or simply exploited a culture where dating is now a pastime, the *search* for intimacy online has become an activity in its own right rather than exclusively a means to find a spouse. Sociologist Zygmunt Bauman discusses this in his work on modern love:

Shopping malls have done a lot to reclassify the labours of survival as entertainment and recreation. What used to be suffered and endured with a large admixture of resentment and repulsion under the intractable pressure of necessity has acquired the seductive powers of a promise of incalculable pleasures without incalculable risks attached. What shopping malls did for the chores of daily survival, internet dating has done for the negotiation of partnership. (2003: 65–66)

The idea of pleasure found in not just dating and hooking but in the act of simply *searching* for someone is identifiable in a range of discussions. A man quoted by sociologists Helene Lawson and Kira Leck illustrates this, framing the online dating pastime as akin to recreational shopping:

> The Internet is a place where people can take risks without consequences. You can experiment with people you wouldn't normally meet or get involved with. You can grocery shop. There are more people to meet. You can play games for a long time. You can look at so many pictures; it's fun like a candy store. (In Lawson and Leck 2006: 195)

Sociologist Eva Illouz explored this idea too and also connected this practice to behaviour usually witnessed in the shopping mall:

> This 'shop-and-choose' outlook is the effect not only of a much wider pool of available partners but also of the pervasion of romantic practices by a consumerist mentality: the belief that one should commit oneself only after a long process of information gathering. (Illouz 1997: 173)

The idea of *the search* being the recreation is underpinned by several key points. Online intimacy-browsing fits into a culture that (a) has already eschewed formal courting rituals, (b) where premarital sex is common and (c) where marriage is being delayed and thus where no burden exists to couple up quickly or permanently.

Something that the internet also facilitates is people engaging in dating – and potentially hooking up with – multiple people simultaneously. While this facilitates betrayal and non-monogamy (Slater 2013b) – and certain websites, in fact, actively trade in helping people meet for cheating purposes (Rosewarne 2015) – it can also be interpreted as evidence of the kind of non-commitment fostered by consumer culture and the ethos that there is always something or *someone* better out there, as Oyer alludes to:

> In the online dating world, this means I know that looking at one more profile creates some chance that this person will turn out

to be the absolute love of my life and will make me happier than
any woman ever could. When I think of it that way, I almost feel
a *responsibility* to go look at another profile. (2014: 7–8)

Key in being able to keep searching and keep postponing a decision
is the sheer volume of site users and possible matches and the simple
fact that users are spoiled for choice; something an online dater quoted
by Slater referenced:

I'm an average-looking guy. I wasn't used to picking and choosing that way. All of a sudden I was going out with one or two very
pretty, ambitious women a week. At first I just thought it was
some kind of weird lucky streak. (2013a: 109)

Alluded to here is the notion of daters having many more options
online – due to the very high number of participants using the platforms – and thus with such a buffet of choice, there is little reason
to hasten a final selection, but instead encouragement to 'shop
around'.

Discussed in this section is the shopping mentality, the buffet
of choice and the ethos that one has to keep searching and swiping
to find perfect fulfilment. As part of this behaviour, a kind of cavalier
attitude has been observed where those met online are dated and
dumped with relative ease.

Disposability

Slater discusses choice as related to online dating, noting that it 'influences mate seekers in at least two ways. First, there's the choice one
confronts at the selection phase. Second, there's the mere existence
of online dating as an easy, discreet mate-finding channel *after* selection has occurred' (2013a: 119). This latter point is particularly important because the very same abundance of options that facilitates a
couple's initial online introduction *remains* after the dyad is established.
In consumer culture, dissatisfaction is constantly encouraged;[3] after
all, if we are content with what we have, we will have few incentives to
return to the market. Thus, within a culture underpinned by the mentality of constant upgrade, it is no surprise that the same methods used
to find intimacy are used to find *more intimacy*. This can, as discussed,
be linked to infidelity, but more simply, just the rapid-fire exiting of
imperfect liaisons in pursuit of something even better.

Something apparent in research on internet communications is
the notion of *weak ties*, an idea explaining that without proximity and

feelings of obligation, relationships formed online are often considered as more disposable because lives are not enmeshed as they would be had a workplace, school or church meeting transpired, and thus exiting such relationships is made much easier (Kraut et al. 1998; Whitty and Joinson 2009; Rosewarne 2018b; Rosewarne 2018c). Such disposability is evident in the experiences of online daters in Barraket and Henry-Waring's research:

> I think the fact that you never have to see them again is wonderful. You know, if it's family or friends, or work or whatever, and it doesn't work out, then you've got all the awkwardness [...] that carries into your life and relationships. (2008: 157)

> I wonder whether it makes everything too easy, so it can lend itself to becoming even more impermanent, relationships even more fragile. It's just too easy to meet people, too easy to cycle through people, so it kind of accelerates the disintegration of long-term relationships, that sort of thing. (2008: 161)

Philosopher Dan Silber presents a similar idea, notably as intertwined with the values of the market:

> Most of us are repelled when we consider the world of dating as a market because doing so seems to threaten all of its participants with objectification. Each individual, with his or her gifts and powers, is simply an exchangeable object that may be traded for another of equal value. (2010: 188)

The idea of lots interchangeable transactions being started and potentially also *ended* online illustrates the idea of the internet thought of as lacking in romance, something alluded to throughout this chapter and a central theme in rhetoric about how the internet has changed intimacy.

Disposability also raises other concerns as related to fully exiting relationships once concluded. Something plaguing all relationships today – not just those instigated online – is the nature of the technology keeping us constantly connected to our former partners. Clean breaks are no longer possible as getting updated on their lives and their new partnerships thus prompting feelings of nostalgia or jealousy is merely a click – or Facebook memory – away (Hill 2019). For those whose partnerships were instigated online, the chance of crossing paths with your match again on another dating platform is also high. For digital natives, this blurring of intimacies is the only lives they have known, but for non-natives, this 'new' reality can be jarring. At the more extreme end, the technology that facilitated the establishment of the union can also be the technology that makes departing from such a relationship dangerous. The cyber-harassment of ex-partners online through to the threat of revenge porn highlights that fully – and safely – disposing of a relationship is something challenged by the internet.

THE UNSERENDIPITOUS, UNROMANTIC MATCH

Part of the rhetoric of the internet's role in the quest for intimacy is that such a method is somehow less romantic than other kinds of meetings. A widespread perception exists that technology is sterile,[4] soulless, cool, calculating and mechanical – what feminist theorist Eileen Boris refers to as 'the antithesis of intimacy' (2010: 13) – and thus to involve it in the search for intimacy extends some of these unromantic qualities to the match found online and, in turn, sully it (Rosewarne 2016a; Rosewarne 2016b).

While perceptions of technology as inhuman constitute part of the explanation for online introductions being construed as unromantic, another key component of this is our cultural ideas about what constitutes romance and a *romantic* love match. Popular ideas about love hinge on several premises – that true love is something predestined and foretold (hence our talk of love being 'in the stars'), and that when the 'stars align', you will find your romantic destiny. Silber explores this concept, where he identified the stranglehold that serendipity has on popular perceptions of 'real love':

> The serendipity of love is important, among other things, because it affords us the illusion of uniqueness. If our love relationships are serendipitous, then they are spontaneous and therefore (improbably) break free of the conditions that otherwise deterministically condition our lives. They are special and meaningful as islands of blissful freedom in a sea of mundane, mechanically ordered events. (Silber 2010: 188)

Film theorist Michele Schreiber makes a similar point, observing how our culture 'consistently celebrates the accidental and fortuitous as being more authentic ways of finding love' (2015: 76).

As argued by almost every romcom or romance novel, there isn't really a place for either technology or, as relevant to this chapter, *the market*, in any true love story (Rosewarne 2016b) that actively soliciting intimacy online is, in fact, somehow at odds with destiny: real love, apparently, should arrive like magic, rather than being worked for, paid for or found online. The undercurrent to this idea is that going online for intimacy is nobody's *ideal* way to meet but, rather, is a kind of *last resort* utilized after offline methods are exhausted. As Aziz Ansari and Eric Klinenberg note in their book *Modern Dating*, a popular perception is that 'using an online site means they were somehow not attractive or desirable enough to meet people through traditional means' (2015: 86).

The supposed lack of romance in an online meet is in fact so culturally entrenched that online dating companies attempt to actively *challenge* it; something discussed by Schreiber:

> Television advertisements for the two most successful American dating sites eharmony.com and match.com actively deflate the negative connotations of online dating's unromantic nature by appropriating aesthetics reminiscent of postfeminist romance films. Eharmony's advertisements are particularly skilful at playing up the serendipitous possibilities available once the site matches you. (2015: 76–77)

Match.com had a commercial that provides a good example of an explicit challenge to the unromantic idea, where it is posited: 'If you're sitting at dinner with Mr. Right, does it matter where you met him?' Here, the online dating company actively attempts to counter the cultural perceptions of online dating as unromantic and, in turn, attempts to counter the well-entrenched notion that the market is at odds with true intimacy.

The way that the internet has revolutionized intimacy-seeking, most notably through the values of the market – notably those pertaining to buying and selling and marketing – has overhauled our intimate lives. Searching and scrolling for an intimate partner has become its own leisure activity, and as explored throughout this chapter, the values of the market have reconceptualized everything from the presentation of self to the belief that there is always something – *someone* – better out there for us.

References

Anderson, Ryan (2016), 'The ugly truth about online dating', *Psychology Today*, 6 September, https://www.psychologytoday.com/au/blog/the-mating-game/201609/the-ugly-truth-about-online-dating. Accessed 11 May 2019.

Ansari, Aziz and Klinenberg, Eric (2015), *Modern Dating*, New York: Penguin.

Barraket, Jo and Henry-Waring, Millsom S. (2008), 'Getting it on(line) sociological perspectives on e-dating', *Journal of Sociology*, 44, pp. 149–65.

Bauman, Zygmunt (2004), *Liquid Love: On the Frailty of Human Bonds*, Malden: Polity Press.

Ben-Ze'ev, Aaron (2004), *Love Online: Emotions on the Internet*, New York: Cambridge University Press.

Boris, Eileen (2010), *Intimate Labors: Cultures, Technologies, and the Politics of Care*, Stanford, CA: Stanford University Press.

Bridges, John C. (2012), *The Illusion of Intimacy: Problems in the World of Online Dating*, Santa Barbara, CA: Praeger.

Cacioppo, John T., Cacioppo, Stephanie, Gonzaga, Gian C., Ogburn, Elizabeth L. and VanderWeele, Tyler J. (2013), 'Marital satisfaction and break-ups differ across on-line and off-line meeting venues', *Psychological & Cognitive Sciences*, 110, pp. 10135–40.

Couch, Danielle, Liamputtong, Pranee and Pitts, Marian (2012), 'What are the real and perceived risks and dangers of online dating? Perspectives from online daters', *Health, Risk & Society*, 14:7–8, pp. 697–714.

De Masi, Sophia (2006), 'Shopping for love: Online dating and the making of a cyber culture of romance', in S. Seidman, N. Fischer and C. Meeks (eds), *Handbook of the New Sexuality Studies*, New York: Routledge, pp. 223–32.

Delaney, Brigid (2009), *This Restless Life: Churning through Love, Work and Travel*, Carlton: Melbourne University Press.

Guadagno, Rosanna E., Okdie, Bradley M. and Kruse, Sara A. (2012), 'Dating deception: Gender, online dating, and exaggerated self-presentation', *Computers in Human Behavior*, 28, pp. 642–47.

Hall, Jeffrey A. (2014), 'First comes social networking, then comes marriage? Characteristics of Americans married 2005–2012 who met through social networking sites', *Cyberpsychology, Behavior and Social Networking*, 17:5, pp. 322–26.

Hill, Katie (2019), 'Facebook memories have the potential to "re-traumatise" you', *10 Daily*, 21 July, https://10daily.com.au/news/tech/a190718kesnq/facebook-memories-have-the-potential-to-re-traumatise-you-20190721. Accessed 19 September 2019.

Illouz, Eva (1997), *Consuming the Romantic Utopia: Love and the Cultural Contradictions of Capitalism*, Berkeley, CA: University of California Press.

Johnson, Ida M. and Sigler, Robert T. (1997), *Forced Sexual Intercourse in Intimate Relationships*, Brookfield, VT: Ashgate.

Kaufmann, Jean-Claude (2012), *Love Online*, Malden, MA: Polity.

Kraut, Robert, Patterson, Michael, Lundmark, Vicki, Kiesler, Sara, Mukopadhyay, Tridas and Scherlis, William (1998), 'Internet paradox: A social technology that reduces social involvement and psychological well-being?', *American Psychologist*, 53:9, pp. 1017–31.

Lawson, Helene M. and Leck, Kira (2006), 'Dynamics of internet dating', *Social Science Computer Review*, 24:2, pp. 89–208.

Mapes, Diane (2006), *How to Date in a Post-Dating World*, Seattle, WA: Sasquatch Books.

Masciotra, David (2015), *Mellencamp: American Troubadour*, Lexington, KY: University Press of Kentucky.

Mathieson, Craig (2019), 'Boogie nights: Are nightclubs still striking a chord?', *The Sydney Morning Herald*, 10 February, https://www.smh.com.au/entertainment/m09cover-20190204-h1atsi.html. Accessed 30 April 2019.

Oyer, Paul (2014), *Everything I Needed to Know About Economics I Learned From Online Dating*, Boston, MA: Harvard Business Review Press.

Putnam, Robert D. (2000), *Bowling Alone: The Collapse and Revival of American Community*, New York: Simon & Schuster.

Roche, Mary M. Doyle (2009), *Children, Consumerism, and the Common Good*, Lanham, MD: Lexington Books.

Rosewarne, Lauren (2011), *Part-Time Perverts: Sex, Pop Culture and Kink Management*, Santa Barbara, CA: Praeger.

Rosewarne, Lauren (2012), 'Love is a (regulatory) battlefield: The ACCC takes on dating website scammers', *The Conversation*, 20 February, https://theconversation.com/love-is-a-regulatory-battlefield-the-accc-takes-on-dating-website-scammers-5377. Accessed 22 September 2019.

Rosewarne, Lauren (2015), 'The hacking of Ashley Madison and the fantasy of infidelity', *ABC The Drum*, 23 July, https://www.abc.net.au/news/2015-07-23/rosewarne-ashley-madison-and-the-fantasy-of-infidelity/6641742. Accessed 17 May 2019.

Rosewarne, Lauren (2016a), *Cyberbullies, Cyberactivists, Cyberpredators: Film, TV, and Internet Stereotypes*, Santa Barbara, CA: ABC-CLIO.

Rosewarne, Lauren (2016b), *Intimacy on the Internet: Media Representations of Online Connections*, New York: Routledge.

Rosewarne, Lauren (2018a), 'Aziz Ansari and the politics of sexual scripts', *Meanjin*, 15 January, https://meanjin.com.au/blog/aziz-ansari-and-the-politics-of-sexual-scripts/. Accessed 17 May 2019.

Rosewarne, Lauren (2018b), 'The online overhaul of courtship', *Pursuit*, 16 November, https://pursuit.unimelb.edu.au/articles/the-online-overhaul-of-courtship. Accessed 18 May 2019.

Rosewarne, Lauren (2018c), 'Facebook dating: Could the tech giant be the ultimate matchmaker?', *ABC News*, 2 May, https://www.abc.net.au/news/2018-05-02/facebook-dating-scary-but-brilliant/9719856. Accessed 18 May 2019.

Ross, Jeffrey Ian (2013), *Encyclopaedia of Street Crime in America*, Thousand Oaks, CA: Sage Publications.

Schreiber, Michele (2015), *American Postfeminist Cinema: Women, Romance and Contemporary Culture*, Edinburgh: Edinburgh University Press.

Senechal, Diana (2011), *Republic of Noise: The Loss of Solitude in Schools and Culture*, Lanham, MD: Rowman and Littlefield.

Silber, Dan (2010), 'How to be yourself in an online world', in K. Miller and M. Clark (eds), *Dating: Philosophy for Everyone: Flirting with Big Ideas*, Malden, MA: John Wiley and Sons, pp. 180–94.

Silber, Dan (2013a), *Love in the Time of Algorithms*, New York: Current.

Silber, Dan (2013b), 'A million first dates: how online dating is threatening monogamy', *The Atlantic*, 311:1, pp. 40–46.

Slater, Dan (2013a), *Love in the Time of Algorithms*, New York: Current.

Slater, Dan (2013b), 'A million first dates: How online dating is threatening monogamy', *The Atlantic*, 311:1, pp. 40–46.

Whitty, Monica T. and Joinson, Adam N. (2009), *Truth, Lies and Trust on the Internet*, New York: Routledge.

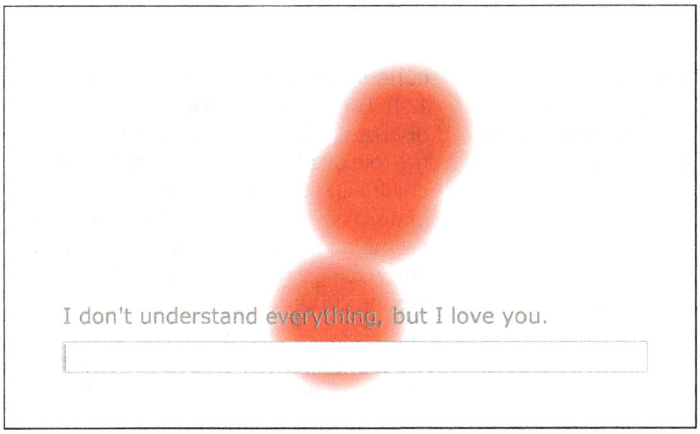

I don't understand everything, but I love you.

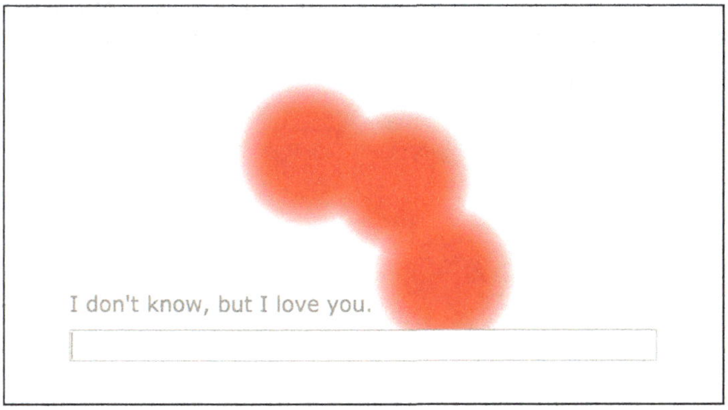

I don't know, but I love you.

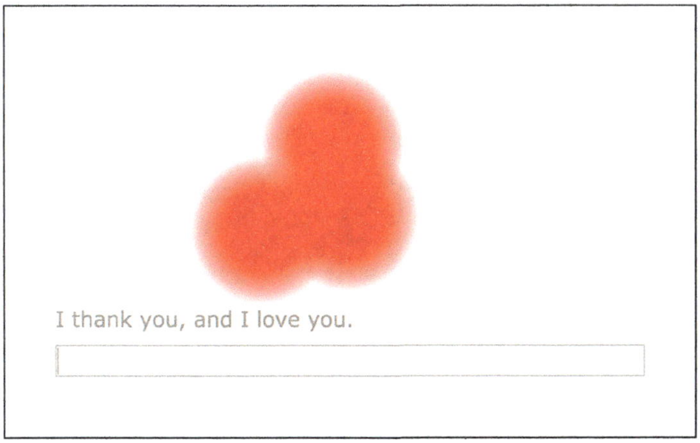

I thank you, and I love you.

FIGURE 1.2–1.4: Antoine Schmitt, *Deep Love*, 2017. Website. Courtesy of the artist and Galerie Charlot, Paris/Tel Aviv.

Notes

1. As Sam Yagan, a cofounder of the dating site OKCupid, notes, 'The only people online in the nineties were socially awkward geeks... So, by definition, they were the bulk of the people doing online dating' (in Slater 2013a: 42).

2. Elsewhere I discuss this idea in the context of teen sexual experiment (Rosewarne 2016a).

3. The criminologist Jeffrey Ross discusses this, noting that, 'A unique feature of contemporary consumer culture is that insatiable desire – the constant demand for more – is now not only normalized but essential to the very survival of the current socioeconomic order' (Ross 2013: 124). The theologian Mary Doyle Roche discusses this same idea in relation to the role of advertising: 'the media keeps children (and adults) in a constant state of dissatisfaction while at the same time promising to relieve that dissatisfaction by introducing a new "must have" product' (Roche 2009: 35).

4. In her article for the British newspaper *The Telegraph*, Sally Brampton identi-fied why she had stopped online dating, 'It seemed to me too sterile, too fast, too lacking in the potential pleasure of love and serendipity' (in Delaney 2009: 38).

2
Love Info-Structures

Since the onset of information technologies, data have been the new language of our social functioning. What we do and how we come across largely depend on data feeds and their profiling. Crucial for that matter is the flexibility of templates with which algorithms let us organize our *self*-portfolios. Equally crucial is the precision of articulating the *selves*, determined by profile-allowed personal categories (name, gender, age, nationality, hobby, etc.), survey formats (e.g. which best describes you: blunt, shy, optimistic, driven), word or character limits (e.g. 150 character bio), image parameters, etc.; they all model the data for browsing and search engines, programmed for effective data flow and distribution.

The obsession with data flow we experience today is perhaps best reflected by autocomplete – the algorithmic search suggestions generated for efficient browsing. Apart from their questionable usefulness and undoubted comic value, search suggestions play a significant role in shaping our discourse of choice and how it affects the way we think about our needs and preferences (and how we formulate them). For instance, Google complete for 'a perfect man is...' would be '... tall'. The change of a subject to 'woman' directs us to 'what is the perfect weight for a woman'. When it comes to love, the poetics of autocomplete enforces and reinforces similar clichés. It either projects the ready-made thinking about romance or reflects recent trends in the thinking development. Autocomplete complicates the search for love by obscuring the search ground. Can we find true love on unsolid predictions of doubtful veracity?

Zach Gage addresses that question in *Glaciers* (2015–16), a series of digital poems on e-ink screens inspired by a Google search complete. The poems are screens-as-wall-clocks connected to the internet, and they grew out of Gage's fascination with predictions generated without human involvement. They also draw on the artist's long-term fascination with large-scale data systems and his attempts to make them more accessible by focusing not on their broad statistical natures but instead on finding their humanity with narrowly tailored queries.

Each 'Glacier' is a unique poem generated via the top three auto-completed results for a specific search query on Google. While they appear static on the wall, each refreshes itself once a day. While this construction process means that the poems may change, because of the immense amount of data flow constructing the most popular autocompletes, it is likely that they will not change for years or even decades. Still, one day you wake up and there is new content. The flow of data in *Glaciers* depicts the liquid meaning of 'solid' phenomena. Love, being one of them, transforms together with the transform-ing language of media that inform its changing senses. Like in Gage's poems:

does he want to.../does she want to...
does he want to marry me
does he want to date me
does he want to kiss me
does she want to be more than friends
does she want to kiss the quiz
does she
i never want to...
i never want to have sex
i never want to see you again
i never want to lose you.
he says...
he says he loves me
he says he misses me
he says I am.

Glaciers

ZACH GAGE

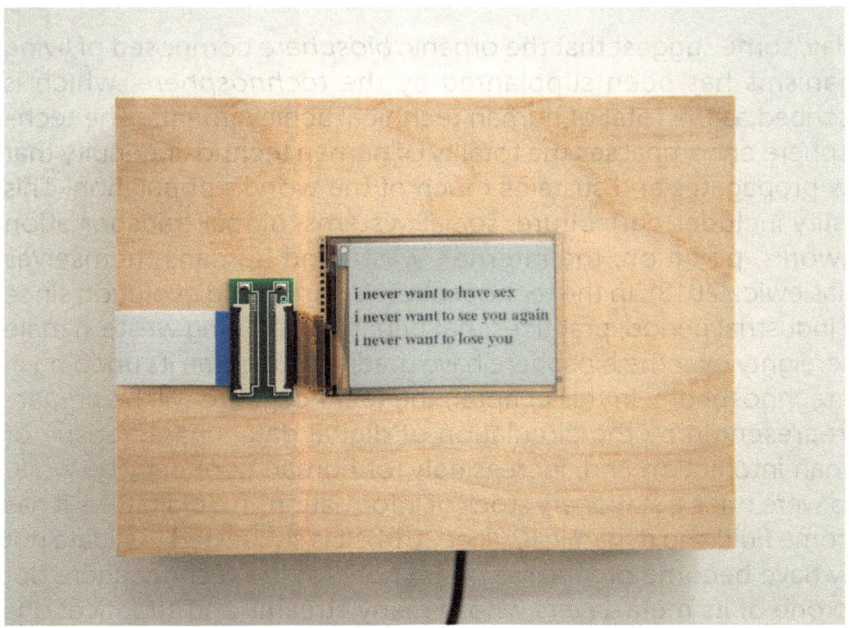

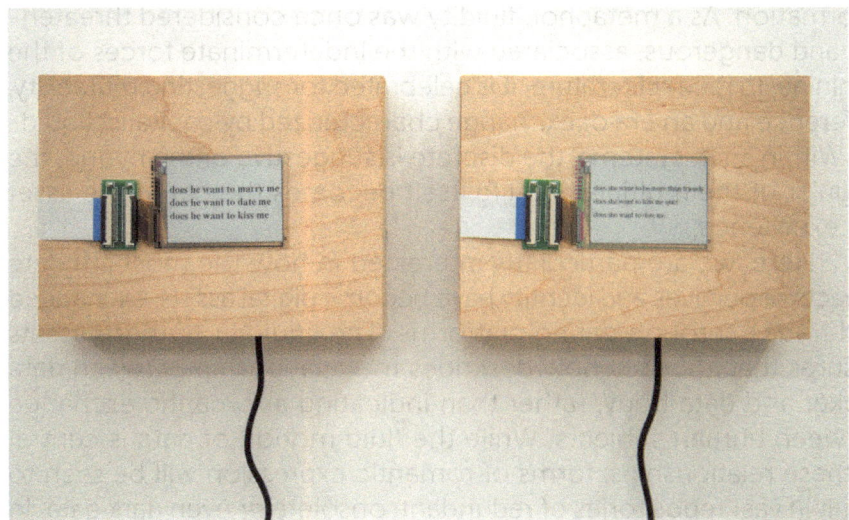

FIGURE 2.1–2.2: Zach Gage, *Glaciers*, 2015/2016. Electronic sculptures, custom wood enclosure, Raspberry Pi, Adafruit pervasive visions 2.7" display kit, ribbon cable extender, edimax 150mbps adaptor, 5.1v 6' microusb power cable, 10ft premium ultraflat cat 6 patch cable, 16gb Sandisk microsd card, 12.7 x 17.8 x 5.1 cm. Courtesy of the artist and Postmasters Gallery, New York.

Romance in a Time of Dark Data

LEE MACKINNON

Today, some suggest that the organic *biosphere* composed of living organisms has been supplanted by the *technosphere*, which is described as the total of human technical achievements. The technosphere encompasses the totality of human techno-ingenuity that now propagates and sustains much of the world's population. This totality includes agriculture, social systems, global transportation networks, pollution, the internet, waste and humans themselves (Zalasiewicz 2018). In the technosphere's ultra-rapid evolution since the industrial period, practices that began by securing western male sovereignty over the biosphere have gradually become its undoing in the technosphere. In this chapter, the technosphere will be primarily represented by the circulation of digital data, characteristic of human interaction and, increasingly, of human connectivity. While data were once a stationary stock of information, in a digital era, it has become fluid and dynamic (Cukier 2014). It is suggested that data not only have become organizing metaphors for the technosphere but also one of its meta-layers whose many streams and stratifications constitute IT systems, eventually congealing into vast *databergs* of information. As a metaphor, fluidity was once considered threatening and dangerous, associated with the indeterminate forces of the feminine. In recent literature, it is celebrated for suggesting mutability, difference and an era of exchange characterized by capitalist liquidity. Within such systems, the discrete is suggestive of the hybrid, the singular of the multiple and life itself becomes the corporate asset par excellence.

Here, we are particularly interested in how the most intimate aspects of our lives and identity have become digital assets, exchanged and traded through dating platforms. The liquidity of these assets ensures that *intimacy* now describes the relationship between data broker and data body, rather than indicating a romantic exchange between human subjects. While the fluid motion of data is central to these relationships, forms of romantic expression will be seen to result in vast repositories of redundant, obsolete or even dark data. In its totality, such data can be seen as the archive and expression of a tenderness not for one another, but towards the current automated paradigm of capitalist appropriation. We will begin by thinking through the metaphorical properties of 'fluidity' and its passage from an ancient signifier of negativity, to a liberatory state, characteristic of data and the technosphere.

GENDERED FLUID

Fluidity is a term that has long been synonymous with a loss of form and has also been used to refer to a loss of meaning. In Ancient Greek literature, the notion of fluidity was at odds with the bounded realities of masculine value and power. Pythagoras famously divided the world into a system of binaries, in which women and men were opposed. On the side of women were formlessness, limitlessness, potential, passivity, darkness and matter, while men were form, light, definite limitation, unity and goodness (Jacobs 1997: 12). These distinctions helped consolidate and naturalize paternal social systems, as described in Anne Carson's examination of women in classical Greece. Being subject to patrilocal marriage, women were defined by their relative mobility that made them the potentially dangerous transgressors of borders, whether physical or geographical. Women's desire itself was considered as fluid and formless as untamed nature, and, as such, a threat to the organization of human societies in the west. Carson explains that ideas regarding women's fluidity are also prevalent in mythology where women often embody the risk of mutability and transgression of limits:

> In myth, woman'[s] boundaries are pliant, porous, mutable. Her power to control them is inadequate, her concern for them unre-liable. Deformation attends her. She swells, she shrinks, she leaks, she is penetrated, she suffers metamorphosis. (2000: 133)

In keeping with their fluid associations, women were described physiologically and psychologically *wet* in Greek philosophy, which Aristotle suggested required the tricky imposition of boundaries upon them (see Carson 2000: 131–2). The marriage and relocation of women were a way to mitigate and manage the threat they posed to the immediate household. At the same time, homosocial relation-ships between men were consolidated through the exchange and sale of women. In short, marriage was a means of relocating women and redeploying the threat of their desirability.

For Carson, love is considered to be the principal motivation for women's flight from form and the immanent dissolution of bound-edness: female desire brings with it the danger of pollution due to its 'liquescent effect' and 'fiery heat' (2000: 134). The language of desire is evoked in terms of waves or floods: it dissolves or melts. It burns. It is the language of liquidity. Today, the language of liquidity will be seen to be liberated through digital networks as endless iterations of data and desire. Indeed, we are used to thinking about the liquidity, not only of desire but also of assets: that is, the readiness of possessions to be converted into flows of cash while retaining a fixed value. In terms of commodities or stocks, liquidity expresses the degree to which a sufficient number of buyers and sellers exist to make prices relatively stable. 'Finance' is the name for such capitalism, understood from the perspective of the investor for whom the convertibility of any asset into cash – its *liquidity* – is distinct from any other utility an asset may have (Meister 2017). It is possible to suggest that women's association with fluidity and formlessness, requiring paternal oversight and 'exchange', helped to naturalize woman's position as a proto-commodity, securing her servitude for centuries.

In recent writing, fluidity has conceptual associations not only with women as objects of exchange but also with data, in an era and economy characterized by the digital database. These states of fluidity are not mutually exclusive, since they are etymologically bound by *matrices*, a reference to the maternal and also to a crucial element of computer science. In early Latin, *Mātrix* referred to a pregnant animal and later came to refer to the terms *womb, source* and *origin*, from *mater* or *mother* (Etymology Online 2020).[1] The technical and figurative sense of the word is derived from the idea of *origin, enclosure* or 'the place or medium where something is developed' (Etymology Online 2020). Matrices also describe rectangular grids of numbers used to describe linear equations – a common tool in electrical engineering and computer science, capable of giving approximations of complex calculations (Hardesty 2013). Matrices are critical to areas such as computer graphics, being that a digital image is composed of rows and columns of pixels, whose colour values correspond to a numerical matrix (Hardesty 2013). The application of the matrix extends to areas such as simulation and genetic analysis. The western etymology of the term *matrix* already seems to chart the place of *origin* away from the biosphere to the technosphere and from Mothers as the originators of life to technocrats who splice genes in the dream of creating motherless reproduction. The life-giving medium that was once the preserve of the maternal animal makes way for the symbolic systems of mathematical and linguistic structures that now code technological existence. Furthermore, in relating women and data, Grinberg claims that in the cultural imaginary, data are like a woman, often evoked by 'the language of liquidity, flows, leaks, streams, oceans, rivers', and indicating that today, data are 'a proto-natural substance that fills objects and bodies' (2017: 112). In equating women with data, each presuppose a neutral medium that is productive, reproductive and can be put to work in the service of patriarchal societies in order to maintain them. In Grinberg's suggestion that data are taking on the status of nature as the medium- or *matrix*-integral to life, we should note that data are increasingly a resource central to capitalist function, becoming one of the main arteries of value circulating through the globally integrated, but unequally distributed, technosphere.

Where data acquire the status of nature, they reproduce notions of uncontrolled matter, awaiting orchestration by the technocrat who can give them a meaning and form. This is parallel with ancient attitudes to women's desire and potential for procreation, which must be contained and controlled by paternal order and exchange. The *natural* resources associated with women in the biosphere are mobilized as new forms of currency in the technosphere. To explain that, I consider that while fluidity has become a metaphor for non-binary inclusivity (among other things), it reiterates operations of power in an attempt to make everything amenable to financial interest and to the homogenization of power.

AUTOFLOW

Fluidity has become characteristic of much more than women, data and finance in the global North and is now used in reference to social systems, identity, sexuality, gender and geographical borders (Lyon 2010; Bayraktar 2015; Sin 2015; Ahmed 2012; Hine 2018; Patent 2017). It seems that the world in its totality can be understood in terms of flows and liquescence. Indeed, gender itself has *become* fluid, rather than describing qualities associated with the feminine. For decades, *the feminine* was a synonym for 'gender', but now, in its redefinition as 'fluid', gender has been untethered from its fetters to become a free-floating signifier that defies easy categorization. It is perhaps in its positive association with non-binary gender that we tend to associate fluidity with a liberatory state, rather than the new protection of old privileges. This positive dimension of all things fluid emerges elsewhere too. For example, Sin draws attention to a sexualized subject that has no essence or fixed identity, being 'fluid' (2015: 414). Patent (2017) suggests that 'borders are historically, socially and culturally fluid', sanctioned only by 'a shared belief in their legitimacy'. Even more compelling here is the understanding of fluidity as a verb that expresses interaction within digital networks. Thus, Zygmunt Bauman has described the digital database as a 'vehicle of mobility' that privileges the global and marginalizes the local (in Lyon 2010: 328). Power, it is asserted, is evaporating from the nation state into the electronically facilitated 'space of flows' (in Lyon 2010: 331). This space of flows points towards the transnational technosphere, where information generated by global communications is the prime commodity. For those U.S. digital corporations who own, organize and trade our data, the fantasy of fluidity is also one of acquiescence to power without physical oppositional form: it is the dream of a seamless horizontal network through which power flows unimpeded.

While we may find room for liberation and liberalism in ideas of fluidity, for writers such as Bayraktar (2015), European ideals of mobility and fluidity are deeply enmeshed with immobility (2015: 4). While Bayraktar refers, in particular, to geographical borders, we can apply such notions to the geopolitical space of the digital, with its promise of fluid interaction and the continuous circulation of free information. For Bauman, the digital database is not only a space of flows but an 'instrument of selection, separation and exclusion' (Lyon 2010: 331). Bodies and borders can be made both visible and permeable by data flows so that they become temporary zones of dematerialization and exception that are more amenable to government control. The term *dematerialization* itself presupposes a physical instantiation that must be actively negated, as does fluidity. This is emphasized poignantly by Ahmed (2012) who notes that present social theory expresses the central motif of fluidity whether in the mobility of Urry or the liquidity of Bauman (2012: 85). Fluidity can thus be a means of obfuscating forms of hegemony and overwriting the bodies of those who oppose it. For example, questioning the pre-eminence of White men in corporate America can be to question the apparent naturalness and structural fluidity of power relations (Ahmed 2017). In this respect, Ahmed has argued that fluidity is an effect dependent upon one's own position: if we come up against a flow contrary to our own, then it becomes a wall that solidly resists our passage (2012: 187). In such cases, one's own body becomes an impediment to fluidity and the smooth functioning of power (Ahmed 2017). An unquestioned submission to fluidity in systems such as digital networks is also potentially a submission to systems of automation. In allowing ourselves to be carried along on such currents, we willingly give rise to (and help to entrain) the automated AI that might soon replace us. Automation and fluidity will be seen to be increasingly conjoined throughout our analysis of dating platforms.

Despite dreams of resistance, the power of digital corporations to draw us into consensual flows and menus of options progressively signifies the automation of the human. As Zalasiewicz acknowledges, we are not directors of the technosphere, but its components, constrained and compelled to keep it in existence for our own survival (2018). While we are the users of digital systems, we are also used by them, not only to entrain automated systems with our aggregated data but also to continually produce data that can serve such ends. The process of human automation is thus evidenced by our use of everyday communication technology. We may become objects of bureaucracy, outsourcing and distribution in a workforce defined by little more than serial components in a consensual machine. Indeed, it has been stated by a number of authors that social media and digital platforms constitute the automation of social relations (Fuchs 2014; van Dijck 2013; Steyerl and Crawford 2017; Lovink 2011). Lovink claims that the 'social' itself has latterly become associated with amorality and individual interest (2011: 6). Attempting to find the 'social' in social media, he questions whether the term has been emptied of its former value by automated structures that entice us with their focus on individual utility. van Dijck claims that social media automate 'engineer and manipulate connections', tracking 'desires by coding the relationships between people, things and ideas into algorithms' (2013: 12). In this respect, she alludes to the way in which desire itself becomes encoded and presided over by automated, algorithmic decision systems. Online dating platforms can be seen to be complicit in this shift toward forms of automation, at the level of both the social and the subject. By automation in dating systems, we refer to the matching, exchange, collation and circulation of human data by algorithms at the behest of other humans. We also refer to the logic of correlation, executed by algorithmic structures that dispense issues of causality. By extension, we can infer that our decision-making is increasingly automated, which is to say, it is more reliant upon structures from which human decision and responsibility have been increasingly eliminated. We explore some of the ways in which this automation is reflected in the translation of human desire into algorithms and liquidity.

BROKEN DATERS

Authors such as Heino et al. (2010) define ways in which the codes and practices associated with romantic love have been radically altered by dating platforms and big data, highlighting aspects of functionality and automation. The authors explore market place metaphors used to describe online dating, highlighting the functionality of the dating website and its evocation of e-commerce sites such as Amazon (2010: 429). In the context of the Amazon warehouse, with its robotic shelving systems and low-paid operatives, we might consider dating a precarious form of emotional labour on zero-hour contracts, where the impetus is on packaging and dispensing as quickly and cheaply as possible. Here, romantic contracts are made and broken through the slightest sign of correspondence with a predetermined calculus of each user. This calculus is established by numerous algorithmic systems, including Q&A forms designed to statistically match likeness. Platform recommendations, or 'matches', are quick-fire in anticipation of pending unlikelihood. We browse profiles of correlated data for a sense of recognition. Arguably, we have rather become the 'servo-mechanism of [our] own extended or repeated image', like Marshall McLuhan's Narcissus, who adapts to his own extension of himself and becomes a closed system (McLuhan 1994: 41). The author reminds us that *narcissus* is taken from 'narcosis' or numbness. Narcissus, we are reminded, has no understanding that he is gazing at his own reflection; as far as he is concerned, he gazes at one who returns his loving gaze (McLuhan 1994: 41).

While daters' behaviour may be gradually 'automated' by the platforms that 'use' them, the automated matching algorithms of dating platforms can be considered to engineer forms of 'homophily' that recall McLuhan's closed systems, where we are increasingly grouped with those most like us (see Chun 2017). The danger with such an insistence on likeness is that difference and otherness become markers of intolerance. Alternatively, we might be constantly surprised by others who have not got the same idea of appropriate romantic behaviour in a game whose conduct has become more differentiated and self-oriented in its outcomes. Today's online ghettos of similarity reflect the limited capacities of non-human agency to assimilate the complex decision-making of human subjects in equally complex social systems. Yet clearly, the digital algorithms that match us by likeness also reproduce existing human bias, ensuring that stratifications of class-based and racialized hierarchy retain their collective force. The assumed disinterest of algorithms in such decision-making masks both the human engineers that develop them and users who buy into them, potentially absolving either of socially divisive engineering.

'Matching algorithms' aggregate data in order to find patterns of behaviour that indicate likeness. Such a method is not interested in causality, but in gathering and statistically analysing data to fit hypotheses established in advance. In this respect, conditions are described in order to achieve certain predetermined outcomes without analyzing the causal determinates of these apparent facts *as* facts. For example, if I see that there is a correlation between those with higher incomes and heterosexuality, I may be well aware that higher income is not a causal factor in determining whether people are heterosexual. Nevertheless, aggregating these datasets may help in establishing certain likelihoods that assist in automated matching. Thus, I perpetuate a social structure in which heterosexuality and higher income have an unquestioned relation. This kind of (il)logic is the logic of the network, which is also the illogic of neoliberal markets. As Chris Anderson stated in his much-cited essay about the *end of scientific theory*, 'This is a world where massive amounts of data and applied mathematics replace every other tool that might be brought to bear'. After all, other means of understanding human behaviour can only ever give us a partial model of otherwise complex systems:

> Forget taxonomy, ontology, and psychology. Who knows why people do what they do? The point is they do it, and we can track and measure it with unprecedented fidelity. With enough data, the numbers speak for themselves... faced with massive data, [the] approach to science – hypothesize, model, test – is becoming obsolete. (Anderson 2008: n.pag.)

While many individuals may still be unaware of the ways in which their participation in online dating platforms reproduces bias, others have already made careers out of aggregating and trading the personal data of online subscribers. An entire industry of data brokers exists to capitalize upon users of dating platforms. For example, the data broking company USDate is registered as a 'General Dating Industry Support Service'. Their website boasts '40 million dating profiles from around the world' and claims that users have agreed on legally binding terms to feature on other dating websites (USDate 2020). Potential buyers for this data are encouraged to populate new dating sites by purchasing existing profiles. It offers bundles of members by country, ethnicity, age, gender and sexual preference. Shoppers are invited to select quantities of pictures per profile. This trade in dating profiles is one of many ways in which the potential relationship between platform members conceals more covert operations. Indeed, the relation established between daters is not *the* relationship per se. The user becomes little more than a series of constellations that help to determine certain demographics by being assimilated into large subsets of likelihoods. The actual relationship then is one established between the singular consumer-dater and data brokers who are being provided with assets that can be turned over and made liquid. In 2017, the NGO *Tactical Tech* and artist Joana Moll, purchased 1 million online dating profiles for 136 Euros from a USDate. The purchase included almost 5 million pictures and personal data including usernames, email addresses, nationality, gender, age, sexual orientation, physical characteristics and personality traits. By analysing the metadata of the supplied images alone, the team was able to trace them to their original online source, demonstrating how insecure our data actually are.

That we become as throw-away as any other obsolete post-upgrade gadget can certainly be existentially terrifying. Equally disturbing is the idea that our seemingly private romantic interactions now provide the fodder for 'data brokers' who buy and sell our profiles, complete with photos and personal data. These data provide a stream of value and a resource for research that no longer belongs to us and whose apparent mundanity is a source of information-rich determinates as data points: our sexual preference, our likelihood of voting for a particular political party, our smile and the way we tilt our head in a photograph indicating bashfulness. These are no longer fond particularities recalled by the singular beloved, but data points extracted by autonomous agents, trained to aggregate further spurious correlations as 'truths'. Thus, despite our relative technical sophistication, we can still exist in a world where visual recognition software is trained to infer traces of homosexual inclination in the distance between subjects' eyes (Tactical Tech n.d.).

Shoshana Zuboff refers to the logic of accumulation in the network as *surveillance capital* (2015). She uses the concept to describe the 'migration of everydayness' into a 'commercialization strategy' (Zuboff 2015: 76). Companies such as Google and Facebook are cited as the arbiters of big data aggregation, which mine data as raw materials:

> Such data are acquired, datafied, abstracted, aggregated, analyzed, packaged, sold, further analyzed and sold again. These data flows have been labelled by technologists as 'data exhaust'. Presumably, once the data are redefined as waste material, their extraction and eventual monetization are less likely to be contested. (Zuboff 2015: 76)

The notion of data exhaust may well make data seem excremental rather than productive. It may also suggest a polluting dimension that attends data output, on a par with those caused by combustion engines in cars that adversely affect human health. After all, it is hypothesized that online activity contributes more to CO_2 emissions than cars do today. A company such as Google processes approximately 47,000 requests per second, which represents an estimated 500 kg of CO_2.[2] Taken collectively, digital services are said to create about 2 per cent of global greenhouse emissions, equivalent to the entire aviation industry (Vaughn 2015). A real-time CO_2 counter named CO2GLE[3] has been created by Joana Moll to highlight the physical infrastructures that constitute our online existence and their power output. Data may feel weightless in their fluidity, yet they constitute part of the 30 trillion tons of materials we use or waste, which is the estimated weight of the technosphere (Zalasiewicz 2018). Leaving a trail of CO_2 and bearing the same weightlessness as the collective human conscience, words of love encircle the globe as data exhaust.

ZUCKERBERGS

The year 2019 marks a moment in which a monument was unveiled to a melted glacier in Iceland. One hundred years ago, the Okjökull glacier covered 15-square kilometres of mountainside in western Iceland: 'Once 50 metres thick, it is now approximately 1 square km of ice less than 15 metres deep and has lost its status as a glacier' (Henley 2019: n.pag.). A commemorative plaque gives a reading of the CO_2 in the atmosphere at the time, making an explicit link to the planetary effects of the technosphere. While qualities of the planet's ice are reduced to water, the fluidity of data begins to congeal to make new mass. In this case, *the databerg* prevails, evoking the temporary triumph of techno-scientific application over the biosphere. Databergs are huge quantities of data that cloy the database, waiting to be categorized or deleted from the non-conscious dimensions of the digital corporation. In recent years, online storage vendors like Veritas, advertise against the *databerg*, likely to cost organizations trillions of dollars. Described as 'a looming data crisis', it is estimated that '33% of data on enterprises can be considered *redundant, obsolete or trivial* [ROT data]' (Siew 2017: n.pag., emphasis added). Companies such as *Arcplace* use the analogy of the iceberg to distinguish between different strata of data value, only a small portion of which is visible 'above the waterline', while around two-thirds of a company's data are 'hidden under the water's surface'. The submerged part of the databerg is subsequently divided into ROT and dark data, which may contain mission-critical information that is unknown (Arcplace 2019). A recent study shows that as much as 80 per cent of data are dark data and only 0.5 per cent are being analysed (Krisifoe 2018).

The figure of the iceberg is an interesting one in a period where ice is melting across the globe at an unprecedented speed and scale. The rate of corporate databerg growth might well be synchronous with the rate at which the icebergs are disappearing, particularly bearing in mind the condition of CO_2 that attends them. The massive databergs of companies like *Facebook*, or *OKCupid*, may well be the only glaciers left to the future. While all else 'melts into air'[4] or is otherwise distributed across the globe as the west's toxic techno-debris, data will reign supreme as 'proto-substance' congealing in every pore and circumstance. In the archaeology of *OKCupid*, our data double may one day be exhumed: our long period on ice will perhaps see us recirculating as data-bait for a dating site that is yet to come. Or perhaps, we will be turned over to the frozen databergs of dark data. This is undoubtedly the fate of all daters: the IT graveyard of the romantic technosphere where the world's love letters, several meters below the visible data-line, gather darkness, remembering the end of human time.

References

Ahmed, Sara (2012), *On Being Included: Racism and Diversity in Institutional Life*, Durham and London: Duke University Press.

Ahmed, Sara (2017), *Living a Feminist Life*, Durham and London: Duke University Press.

Anderson, Chris (2008), 'The end of theory: The data deluge makes the scientific method obsolete', *WIRED*, 23 June, https://www.wired.com/2008/06/pb-theory. Accessed 10 February 2020.

Arcplace (2019), 'Data analysis and classification: The Databerg', https://www.arcplace.ch/en/offer/solutions/data-analysis. Accessed 16 December 2019.

Bayraktar, Nilgun (2015), *Mobility and Migration in Film and Moving Image Art: Cinema Beyond Europe*, London: Routledge.

Carson, Anne (2000), *Men in the Off Hours*, London: Jonathan Cape.

Chun, Wendy H. K. (2017), 'We are all living in virtually gated communities and our real-life relationships are suffering', *WIRED*, https://www.wired.co.uk/article/virtual-segregation-narrows-our-real-life-relationships. Accessed 10 February 2020.

Cukier, Kenneth (2014), 'Big data is better data', *TED Salon Berlin*, https://www.ted.com/talks/kenneth_cukier_big_data_is_better_data/transcript#t-241600. Accessed 10 February 2020.

Etymology Online (2020), 'Matrix', https://www.etymonline.com/word/matrix. Accessed 10 February 2020.

Fuchs, Christian (2014), *Social Media: A Critical Introduction*, London and New York: Sage Publishing.

Grinberg, Yuliya (2017), 'The emperor's new clothes: Implications of nudity as a racialized and gendered metaphor in discourse on personal data', in J. Daniels, K. Gregory and T. McMillan Cottom (eds), *Digital Sociologies*, Bristol and Chicago: Polity Press.

Hardesty, Larry (2013), 'Explained: Matrices', *MIT News*, 6 December, http://news.mit.edu/2013/explained-matrices-1206. Accessed 27 February 2020.

Hayles, N. Katherine (1999), *How We Became Posthuman: Virtual Bodies in Cybernetics, Literature and Informatics*, Chicago and London: University of Chicago Press.

Heino, Rebecca D., Ellison, Nicole B. and Gibbs, Jennifer L. (2010), 'Relationshopping: Investigating the market metaphor in online dating', *Journal of Social and Personal Relationships*, 27:4, pp. 427–47.

Henley, Jon (2019), 'Icelandic memorial warns future: Only you know if we saved glaciers', *The Guardian*, 22 July, https://www.theguardian.com/environment/2019/jul/22/memorial-to-mark-icelandic-glacier-lost-to-climate-crisis. Accessed 1 December 2019.

InternetLiveStats.com (n.d.), 'Google searches in one second', https://www.internetlivestats.com/one-second/#-google-band. Accessed 10 February 2020.

Jacobs, Frederika H. (1997), *Defining the Renaissance Virtuosa: Women Artists and the Language of Art History and Criticism*, Cambridge: Cambridge University Press.

Krisifoe, Bella (2018), 'Marketing in the dark: Dark data', *CMOs Studio IBM*, https://www.ibm.com/blogs/think/be-en/2018/04/24/marketing-dark-dark-data/. Accessed 1 December 2019.

Lovink, Geert (2011), *Networks without a Cause: A Critique of Social Media*, Cambridge: Polity Press.

Lyon, David (2010), 'Liquid surveillance: The contribution of Zygmunt Bauman to surveillance studies', *International Political Sociology*, 4:4, pp. 325–38.

Marx, Karl and Engels, Friedrich (2008), *The Communist Manifesto*, Oxford: Oxford University Press.

McLuhan, Marshall (1994), *Understanding Media: The Extensions of Man*, Cambridge, MA and London: MIT Press.

Meister, Robert (2017), 'Reinventing Marx for an age of finance', *Postmodern Culture*, 27:2, https://muse.jhu.edu/article/680224. Accessed 1 February 2020.

Moll, Joana and Tactical Tech (2018), 'The dating brokers: An autopsy of online love', https://datadating.tacticaltech.org/viz. Accessed 1 December 2019.

Moll, Joana and Tactica Tech (2019), '*CO2GLE*', http://www.janavirgin.com/CO2/CO2GLE_about.html. Accessed 1 December 2019.

Patent, Volker (2017), 'What are borders?', *Open Learn, The Open University*, https://www.open.edu/openlearn/society-politics-law/geography/what-are-borders. Accessed 5 October 2019.

Quito, Anne (2018), 'Every Google search results in CO2 emissions: This real-time data viz shows how much', *Quartz*, 7 May, https://qz.com/1267709/every-google-search-results-in-co2-emissions-this-real-time-dataviz-shows-how-much/. Accessed 21 July 2019.

Siew, Alfred (2017), 'Veritas: All that useless data will cost companies trillions of dollars', *Techgoondu*, 17 March, https://www.techngoondu.com/2016/03/17/veritas-much-useless-data-costing-companies-trillions-dollars/. Accessed 1 December 2019.

Sin, Ray (2015), 'Does sexual fluidity challenge sexual binaries? The case of bisexual immigrants from 1967–2012', *Sexualities*, 18:4, pp. 413–37.

Steyerl, Hito and Crawford, Kate (2017), 'Data streams', *The New Inquiry*, 23 January, https://thenewinquiry.com/data-streams/. Accessed 1 March 2020.

Tactical Tech (n.d.), 'Quantifying homosexuality: A critique', https://ourdataourselves.tacticaltech.org/posts/40-quantifying-homosexuality-critique. Accessed 1 March 2020.

Therkelsen-Terry, Nick (2017), 'Why Watson? More accuracy. Less training time. Insanely better CX results', *ibm.com*, https://www.ibm.com/blogs/watson/2017/09/more-accuracy-less-training-time-better-cx-results-with-watson/). Accessed 21 July 2019.

USDate (2020), 'Dating business support services', https://www.usdate.org. Accessed 1 February 2020.

van Dijck, José (2013), *The Culture of Connectivity: Critical History of Social Media*, Oxford: Oxford University Press.

Vaughn, Adam (2015), 'How viral cat videos are warming the planet', *Our World*, https://ourworld.unu.edu/en/how-viral-cat-videos-are-warming-the-planet. Accessed 15 October 2019.

Virilio, Paul (2007), *Negative Horizon: An Essay in Dromoscopy*, London and New York: Continuum.

Zalasiewicz, Jan (2018), 'The unbearable burden of the technosphere', *UNESCO*, https://en.unesco.org/courier/2018-2/unbearable-burden-technosphere. Accessed 1 March 2020.

Zuboff, Shoshana (2015), 'Big other: Surveillance capitalism and the prospects of an information civilization', *Journal of Information Technology*, 30:1, pp.75–89.

Notes

1. This shift from animal to human bearer of life and load reminds me of McLuhan (1994: 93) and Virilio's (2007: 40) respective observations that 'the first pack [bearing] animal was woman'.

2. https://www.internetlivestats.com/ one-second/#google-band. Accessed 14 May 2021.

3. On average, a visit to the sites emits 0.037 g of CO_2: http://www.janavirgin. com/CO2/CO2GLE_about.html. Accessed 10 April 2020.

4. A reference to Marx and Engels who, in 1848, suggest that 'all fixed, fast-frozen relations [...] are swept away, all new-formed ones become antiquated before they can ossify. All that is solid melts into air [...] and man is at last compelled to face with sober senses his real conditions of life, and his relations with his kind' (2008: 6). Although they are speaking about the bourgeoisie continually revolutionizing the tools and relations of labour, the paragraph has great resonance with the climate crisis today.

3
Mediated Matchmaking

Out of the many definitions of *mediation*, the most convincing and inclusive seems to be: mediation is magic. If time-space travel still appears a quintessential mystery of nature, mediation unravels it by taking us beyond the limits of physical laws. 'Mediated contacts' hack our temporal and spatial perceptions. They also warp the relations between distance and time – their geopolitical and socio-cultural dimensions. A Skype across the Atlantic is not a mere compression of time, but the flattening of 'different times' within a single moment of 'shared universal time'.

Media operations across different spaces and temporalities bring out previously unrealized complexities around distance, time and communication practices. Many of them are mirrored in the construction and operation of *A Truly Magical Moment* – a 2016 kinetic installation by Adam Basanta that brings into play people, media devices and communication protocols – all embedded the cultural tropes of a romantic encounter. *A Truly Magical Moment* takes its origin in a period when Basanta and his girlfriend were separated by travels, and when many of the artist's friends stayed in long-distance relationships. Thinking about tools for those kinds of relationships provoked questions about changes in couple intimacy and the intimacy's possible technological reproductions. The installation emulates romantic mediated interactions in a choreography of two iPhones adjusted to diagonally positioned selfie sticks that protrude from a rotating platform. The iPhones face each other, and as two participants log in to the installation's system, they perform a romantic face-to-face dance: 'the lovers dance' – one you can see in *Titanic*. The iterative and layered cultural framing of this movement sequence (or 'dance') as means in which we can fall in love, or a way of representing a non-verbal performance of love, is both central to understanding the love code embedded in the work, shared by a large number of people born after 1980. Basanta's work translates that code into the media language of digital natives to be reproduced in technological environments, which the code now invariably inhabits.

The installation involves a multitude of overlapping physical and virtual spaces: the space of the phone, the space of a given physical location, the space of a time zone and the space of a mediated interaction, which is a space between the mediated presences of the mediated parties. There is also the space of the network connection – represented by wave transmissions, physical cabling, satellite dishes and satellites and, of course, the physical space between these elements itself. This includes the connection's various fluctuations in speed and quality, which can itself be envisioned as material space.

Finally, there is the notion of a digitalized movement of a body or a body moving in the digital realm. Even if mediation offers certain (disembodied) possibilities with regard to time and space, it cannot fully replace the physical movement and its potentialities. The main effect of tele-presence is ironically that of absence. A telematic image only reinforces the lack of actual bodily presence in the space. The body cannot be visually represented through digital technologies. Representation always implies an absence of the 'Real Thing'. Perhaps, this impression is generational, but the sentiment imbued in *A Truly Magical Moment* enunciates the obscurity of the mediated experience. The moment that the work renders is far from being *truly magical*. All the contingent side effects of online communication, the aesthetic markers that are conveniently left out of commercial technological narratives – occasional deterioration of quality, the microphone going out for a moment, disconnection from WiFi and even the screen freezing completely – disturb the authenticity of interactions that mediation tries to create. We have accepted a lifestyle in which seemingly magical abilities (like bridging or connecting different spaces and times through video chat) result in the absence of magic, a pale imitation of connection. Perhaps, the effect created by absence generates a more organic response than a digital 'presence' ever could.

A Truly Magical Moment

ADAM BASANTA

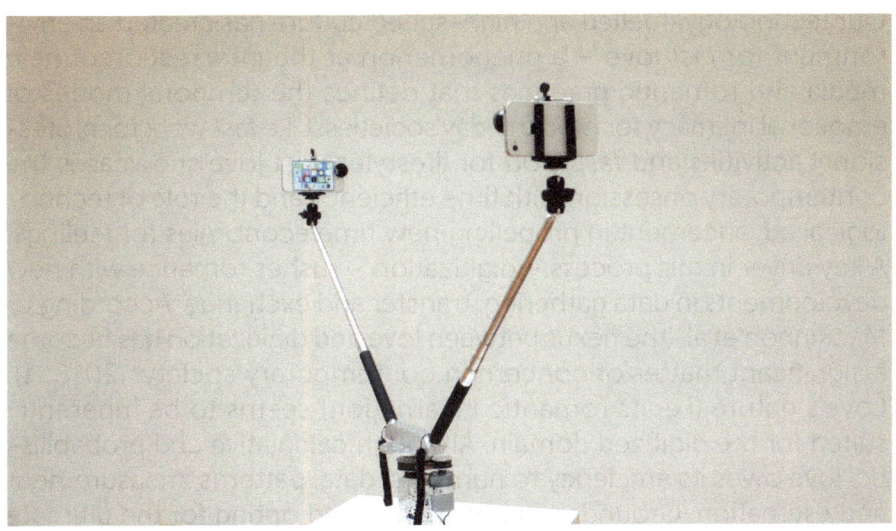

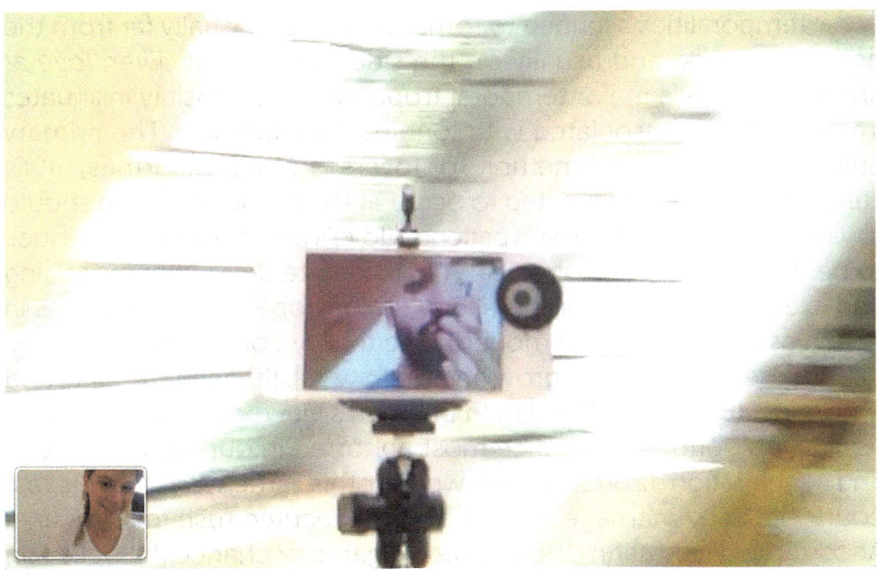

FIGURE 3.1–3.2: Adam Basanta, *A Truly Magical Moment*, 2016. Interactive kinetic sculpture, 2 iPhones 4S, selfie sticks, aluminium, electronics, Bluetooth chips, FaceTime video chat software, 1 m x 1 m x 1 m. Courtesy of the artist.

Fast Love: Temporalities of Digitized Togetherness

ANIA MALINOWSKA

Our technology-fuelled and high-speed culture has created an environment for *fast love* – a phenomenon at the intersection of new media and romantic practices that defines the temporal modes of emotional intimacy for modern-day societies. Like *fast work* for professional activities and *fast food* for lifestyles, fast love showcases the contemporary obsession with time efficiency and the role of technological advancement in propelling new time economies for feelings. A key driver in this process – digitization – rushes romance with new developments in data gathering, transfer and exchange. According to MacKinnon et al. 'the nexus between love and digitization has become a significant matter of concern in contemporary society' (2018: 1). Love's nature (i.e. its romantic incarnation) seems to be inherently suited for the digitized domain. Although calculative and probabilistic, love owes its efficiency to numbers, data, patterns, measurement and estimation. Grounded in impatience and opting for the ultimate appositeness (at a minimum waste), love is perfectly economical and disciplined, especially in terms of time.

Temporalities ascribed to romantic love are usually far from the notion of speed and usually on the side of slowness. Even *love at first sight* – love's major temporal trope, which ostensibly insinuates immediacy – is associated with longing and duration. The primary culturally sanctioned time unit for lovers is waiting (Barthes 1990; Lahad 2017). Love is expected to occur at the proper time and should always entail anticipation with patience (love's primary temporal virtue), to be deserved in the first place. A considerate delay in approaching and executing relationships makes a major trope of romance – as in fiction as in real life. Most modern love theory convinces that 'desire needs time to germinate, grow and mellow [...]. [It] needs tending and grooming' (Bauman 2003: 11). In Bauman's view, 'a delay of satisfaction [is] no doubt the sacrifice most abhorred in our world of speed and acceleration' (2003: 12). No wonder this same world cheers fast courtship, early marriage and otherwise executed rush for coupling. According to Luhmann (1986), love is a game of chance that relies on quick reflexes. Romance is a 'matter of time' for which 'first come, first served' is not so much a dictum of digital protocols but an imperative of culturally cultivated paces of romantic praxis that those protocols invariably enhance.

The trajectory of change in 'distributing love's codes and behaviours through the social systems' – from analogue, temporal, and calculable towards digital, ultra-rapid and computational (MacKinnon 2016a, 2016b) – reveals a specific construction of romantic urgency; this urgency emerges from and depends on operational convenience inscribed in love codes – now additionally boosted by digitization. Accessibility and manageability are considered major factors for romantic interactions, prioritizing immediacy and making romantic endeavours 'a calculation of chance' (MacKinnon 2016a, 2016b), based on *the calculation of time* under *the calculation of convenience*. In this way, modern times amorous encounters are time-flexible exchanges of romantic/erotic codes where the reliance on profiling and data (e.g. *online dating*) as well as programming and algorithms (e.g. *sex bots*) ensures zero-risk interactions (Badiou 2012; Horvat 2015; Han 2012) across 'the variety of times' (Urry 2008: 180) and 'the ever-accelerating contraction of duration' (Schweizer 2008: 6).

MIXED (TEMPORALITIES OF) FEELINGS. CONTEXTUALIZING FAST LOVE

In a personal account of a mediated break-up that happened over a mobile and was followed by the arrival of a love letter, dispatched by post before the relationships' abrupt termination, Lee MacKinnon (2016a) describes the clash of two discourse machines: the temporal and calculable machine of the analogue and the ultra-rapid computational machine of the digital. The clash unveils 'different technical systems of communication and their ability to execute our decisions' (MacKinnon 2016a). This also shows how information exchange (in that case, the exchange of feelings) can be suspended between two regimes – the older one that she defines as 'probabilistic and determining', and the more recent one, 'where temporal and spatial relations are expedited by digital processing' (MacKinnon 2016a). Although MacKinnon describes these regimes in terms of their executive manner – literal/predigital vs. computational/post-digital – she in fact speaks of temporality and two temporal planes, defined as 'real time' and 'digital time', that are now in use for experiencing intimacy and togetherness.

Those two temporal planes organize our thinking about romantic relationships in terms of opposites: flow/fast, analogue/digital and natural/artificial. The *slow* (analogue)/*fast* (digital) dyad signifies the supposed shift from love understood as a steadily developing attachment (i.e. happening at a right time and proper duration) towards love that became a frenetic and rapidly satisfied urge. The other dyad, analogue (natural) and 'technologically handled' (artificial), supports a claim that the acceleration of romantic experience – brought on by the progression of technology – disintegrates the event of love. In effect, the love we practice today appears to be something inauthentic and unnatural. It also appears to be something broodingly new. But as Benjamin Bratton (2016) observes, '[the] lazy association of analogue systems with physics and nature, and digital systems with artifice and artificiality dulls and confuses our debates about technology'. It similarly dulls and confuses the debates on our experiences with technology, imbuing everything we do with a false sense of novelty.

Love under technology is a part of a bigger social process whereby technological modifications of lifestyles and the tempo of living substantially alter 'practices and actions [...] and associational structures and patterns of relationship' (Rosa 2015: 74). But those alterations do not affect the universal function those practices entail. Nor do they change their social and cultural significations. If we agree with Luhmann (1986) that romantic love is primarily a code deployed across the social system by means of 'semantic devices' (e.g. 'I love you'), we understand that the major change from its traditional form to its modern incarnation relates to 'various symbolically generalized media of communication' (Luhmann 1986: 5). Also, if we agree that – 'despite all the emphasis on [...] passion' – love is actually 'a model of behaviour that could be acted out and which one had in full view before embarking on the search for love' (Luhmann 1986: 20), we realize that the technological intervention means simply a transfer of the existing codes of romance onto the digital media of communication (or vice versa).

What has changed, however, is the experience of time around love codes. Temporal dimensions (e.g. *online time*) and agencies (e.g. *machine time*) that followed the advent of digital technologies have significantly intensified the realities of social time (e.g. *institutional time*) and cultural time (e.g. calendars) we operate by. This has frustrated our understanding of time as something linear, cyclic and successive, bending the human temporal capacity (and flexibility) in ways that strain our senses and operational spans. As Baker observes,

> when we interact with digital technology, we interact across multiple temporal rhythms. The time of the user meshes with the time of the machine, including a synchronous time of the software, the non-sequential time of the database, the time of the network and the time of the other users. In general, the multi-temporality of the digital presents an alteration to the way we experience the occasions and events of our everyday lives, beyond a chronological sequence of events. (2012: 14)

'Acceleration of feeling' that we have experienced for quite some time is an effect of the merging of multiple temporal realities. Like other types of encounters, love takes place in environments that inflict a variety of paces, followed by new modes of interactions (synchronic long-distance presence, mediated communication, etc.). Those inter-actions are not sequential (do not happen one by one) but accumulative (they happen all at once). Hence, when we talk about fast love, we do not talk about a transition from analogue to digital – from slow to fast, from calculable to computational, from temporary to ultra-rapid – but about the build-up of all these modes and their clash in a romantic encounter.

FAST LOVE

When we open a WhatsApp to text 'I <3 U', we engage in the process of fast love. When we reach for an online-dating service or hang out romantically via an immersive VR sets (with a hologram or a human significant other), we also practice fast loving. Pigeon post, as used for romantic purposes, can also be considered an instance of fast loving or at least its predecessor or a stage in its development.

I define fast love as a temporal mode for practicing romance that emerged alongside human reliance on technologies and technological advancement. I also define it as a marker of the changing perception of love practices and love's temporal nature associated with what has been identified as social acceleration. Although fast love is linked with the high modern pace of living, it diverts from the critically justified understanding of social acceleration as a transition from pre-digital to digitalized (i.e. from analogue to computational). Rather, it denotes an accumulation of various temporal planes in the contemporary prac-tice of everyday life – in that matter, the practice of romance. There-fore, unlike other *fast* phenomena (fast travel, for instance), fast love is not intended as an antithetical term, i.e. something that opposes *slow love*. It rather expresses a temporal logic – reflected in the nature and history of modern romance as a social practice – that anchors think-ing about the contemporary pace of romantic encounters outside 'the well-worn question "What is time?" asking instead: "How is time produced?"' (Baker 2012: 2).

The production of time for romance situates fast love within the social economies of time efficiency and time flexibility. At the same time, it signalizes the clash between the social (or market) idea of time-making and the private (non-social) motivation for making time for love. Libidinal economies, although they operate by the 'the [market] logic of exchange, accumulation, and profit' (Pettman 2009: 26–27), negotiate the market economy in how they deploy pleasure and desire. They specifically bring desire back to the individual, taking it out of the grips of market and social productivity. The production of time for romance decelerates market imperative in how it accelerates personal pleasure. In other words, whenever we make time for love, we neglect the imperative of making time for work. Approached that way, fast love might be seen as a form of exploiting market temporality (and social acceleration), a form that (by making time for love) puts pleasure over capital.

Of course, media technologies employed in the process anchor fast love deep in the market economy, making time-production for love a way less romantic. As Solange V. Manche (2019) observes, dating apps make us 'undeniably part of a shared temporal financial milieu: [they allow us] to function as the perfect employee[s] in a *liquid* market [by making us] choose to have sex at moments that do not hamper [us] as neoliberal being[s]'. Without doubt, dating apps are a new modality of desire. According to market research, the very business is worth 2.5 billion dollars. Still, the attempt to link desire back with the production of time for individual pleasure or fantasy (rather than social or market benefit) is there. And when enacted, it carries some subversive potential (with little effect though).

Two nexuses that organize fast love in terms of efficiency and digital structuring are *the datafied constant* and *the digital instant*. *The datafied constant* represents the outcomes of technologically managed relationships, which owe their success to well-profiled fast information. *The digital instant* signalizes the role of media-enabled immediacy for an encounter whereby technologically granted accessibility – of people, romance hubs, intimacy software and hardware – is a crucial, yet problematic, element of simulating presence and proximity in ever de-socializing modern societies.

The *datafied constant* brings to the fore technological preoccupation with collecting, sorting, combining and otherwise processing romance-related data. It also brings to the fore the role of data in the initiation, maintenance and control of a successful match – something that modern societies have been obsessed with due to the social significance of matrimony and its religious, political and economic function. As confirmed by a number of research, online relationships show a high rate of stability and often happen to continue longer that offline arranged marriages. This makes the technologically managed romance a new neoliberal ideal. Philosophers Alain Badiou (2012), Byung-Chul Han (2012) or Srećko Horvat (2016) accuse online dating platforms for promoting low-risk and effortless romance; 'love without suffering', as

Badiou calls if after a Meetic ad, subscribes to the narcissistic culture of no negativity, preoccupied with well-groomed self-presentation, accurate estimations of options and maximization of positive results. Illouz (2007) links this calculation of risk in romance with a transfer of capitalist strategies onto emotional processes – a phenomenon she terms *emotional capitalism* and considers a corruption of love processes. Indeed, information from datasets, device protocols and interfaces grant dating parties a great sense of control over the romantic experience. Keeping a track of media status (e.g. 'last seen') or the actual monitoring of partner's whereabouts (geolocation apps) are only a tiny fraction of common romantic surveillance practices used to minimalize failure of a relationship (Thylstrup and Veel 2018).

The *digital instant* signalizes two things: one, the efficiency inscribed in mediate encounters (as we meet online, we can access anybody from anywhere in the world in what feel like 'real time'); two, the clash of human and technologically generated temporalities – especially in terms of instantaneity and continuity. While the first is essential for contemporary relationships, the other seems deeply problematic. Media objects are never permanent, but are constantly developing forms, 'produced through an ongoing technological process' (Baker 2012: 5). Digital images, sounds, text messages, 'weather static or in motion' are never stable because they are 'a result of continuous and ongoing computations' (Baker 2012: 5). The environments they co-create 'do not exist' but are 'continually *in the making*' (Baker 2012: 5). So are the data those environments produce. The material and temporal reality of media objects, even if able to emulate the human experience, is far from the organic reality that people live by. In other words, mediation creates illusions of time and presence in which human and technological immediacies (in the case of the latter, digital) negotiate the senses of realness and authenticity for human and machinic experience alike.

Elements of fast love that explain the nature of those negotiations are fast match, fast medium and fast subject, which in themselves outline the evolution of love's technic towards digitalism.

Fast match

Finding an ideal partner in a relatively short period is one of romance's main ambitions. Analogically, impatience is one of romantic love's prime cultural units of time. Modern societies have developed a number of methods for fast couple-matching, including traditional ways such as arranged marriages or 'professional' matchmaking. The most distinctive strategy directly associated with modern societies and the acceleration of life tempo is speed-dating: publicly arranged meetings that allowed an exposition to a large number of potential partners in a short time that exploded in late 1980s. Speed-dating was organized in a form of quick face-to-face conversations, on the basis of which the participants decided on the fate of a match. As a method based on data analysis, speed-dating may be considered a predecessor of contemporary love databases for quick estimations that inspired the idea of pre-determined information processing for a match. Of course, unlike current algorithmic date hubs, the *predicted outcome value* of speed-dating depended on human-to-human calculations conditioned by interactional immediacy (Houser et al. 2008). Nevertheless, the time-control and time-saving methods speed-dating initiated are still with us and work as a model of romantic effectiveness (Hollander and Turowetz 2013).

Online dating, which took over in the 1990s and which transferred the initiation of romance into the internet, replicates the temporal premises of speed-dating basing match formation on informational precision. In a study assessing the effectiveness of online dating for continued offline interaction, Sharabi and Dykstra-DeVette (2019) point out that the templates for coordinating online romantic inter-actions rely on meticulous protocols of self-presentation that make the trajectory from first email messaging (as an initial point) to the first offline encounter (the final point of the matchmaking process) most possibly quick and productive. They also stress the increasing willing-ness to rely on the veracity of machinic computation that, according to MacKinnon, signalizes a shift 'in favour of immediately quantifiable coordinates' (2018b: 163). In her analysis of the role of the GS match-ing algorithm for non-erroneous pairing, MacKinnon highlights the primacy of protocols over subjectivity for the immediacy of profiling and selection. As she observes,

> [i]n the context of online dating platforms, the potential lover becomes a list of discrete *menu's* – increasingly informational and calculable, considered in terms of *user's* ability *to control/command/alternate/delete.* Human attributes can be mapped on to the technical devices, whereby the potential partner is assem-bled according to techniques associated with digital processing: editing, construction choice, convenience, ubiquity, obsoles-cence, discretisation – features associated with digital technol-ogy and its protocols. Here, speed may be associated with the elision of meaningful translation between one and the other that can ameliorate desire only by eliding the threat of any gap with the immediacy of a new object or 'gadget.' (MacKinnon 2018b: 163)

Despite the often-lamented superficiality of online datasets (Epstein 2007), the avid interest in online dating services confirms a growing preference for codified and computed love matches. Our trust in algorithmically controlled matches is especially reflected in the user statistics of internet romance services (Finkel et al. 2012). Piskorski informs that 1,804,993 users (1,493,205 of whom are registered as American citizens) used the OKCupid site between 1 October 2010 and 15 December 2010. A similar number of 1,804,933 singed up or logging users were registered for eHarmony during exactly the same period (Piskorski 2014: 26). Although motivations for this massive reliance on computers with the matters of the heart vary, we do tend to delegate choice and decision-making tasks to machines with more trust. Frischmann and Selinger (2018) connect it with the disappointment in human reasoning that we consider less 'rational' in terms of efficiency, goals, utility and the proper assessment of benefits. According to Oscar H. Gandy Jr., since the discovery of machine intelligence, our approach to rationality became qualitatively bipolar; we started to distinguish between the rationality of *idealized intelligence*, produced by 'a difference engine that engages in rapid computation without errors in calculation, and more critically, without any systematic bias introduced by emotional distractions', and 'the sometimes slow, sometimes fast, error prone, easily distracted, and routinely distorted information processing by humans' (in Frischmann and Selinger 2018: 192–93). As the reliance on machine intelligence embedded in the media devices we use daily becomes a standard for reasoning about love, the role of those devices shifts from mere mediators to affective agents. The modern media take over many bits of emotional labour; with this, they accelerate the dynamics of our romantic conduct and its inherent protocols.

Fast medium

From messenger to Messenger, the history of modern romance proves that love is bound to solutions for fast communication. When electrical telegram was introduced for common use in the nineteenth century, at least one-third of communiques sent over the wires were the messages of love (Bruton 2015; Rosenkrantz 2003). In *Alone Together: Why We Expect More from Technology and Less From Each Other* (2011), Sherry Turkle speaks of *temporal control* as a crucial factor for constructing relationships today. Her study on the use of media in romantic engagements explains specifically how new technologies have granted relationships a new sense of maintenance and continuation (2011: 190).

The major advantage of 'loving with media' is the elimination of waiting ('They send an e-mail, and they expect something back fast', Turkle 2011: 166). Another is the acceleration of romance. The 2014 study by Storey and McDonald describes the effects of texting on 'speeding up the development of their romantic relationships' (118). Many of their interviewees admit that it significantly enabled their romantic performance, especially in terms of speeding up intimacy:

> 'I think [texting] speeds things up more than anything, because now [...] you can constantly be in contact'. [...] 'I think I'm closer to him because you get to know someone quicker 'cos you're texting them and like we do text quite a bit. And like in the early stages of us getting together that's kind of like how we got to know each other and like we were texting quite a bit and so I think it does help you get to know them a bit closer'. [...] 'I think you can be a bit more risky, a bit more rude, a bit more cheeky'. [...] 'I think it's easier for people to let themselves get more intimate than what it previously would've been'. [...] 'Yeah, I think especially when you're getting to know them, it's easier to be a bit more brash than say if you just met them on the street out of the blue. I don't think you'd be like, huh [he makes a noise suggesting sexual excitement]'. [...] 'Yeah, you've got more confidence to message each other haven't you. Rather than face to face.' [...] 'I try to act the same in texts as I would do in person, but then I think that you do find yourself talking on text, or in fact on Facebook chat, you find yourself saying things that you probably truly wouldn't say in person'. [...] '[T]exting: helped us seduce each other.... It allowed us to express ourselves and say those things which made us feel the urge and need for the other person even more.... It was precisely through text messages ... that very "romantic" and breath-taking things were said between us'. (In Storey and McDonald 2014: 118–19)

But the function of media devices for romance exceeds the mere exchange of content; most software, hardware and interfaces actively participate in co-creating the protocols of love. Many common forms of intimacy would not be possible without modern communicators and their specific design. Cybersex, for instance, owes its entire idea to the evolution of VR equipment, wearables and teledildonics. Much effort in this respect has been inspired by the need to overcome physical remoteness. Interesting how geographical and social distancing impels technological advancement in terms of the transmission of body contact and exchange of intimacy. Even if still primitive, devices to emulate physical interactions – such as Kissinger (a mobile add-on specializing in the transfer of kisses) or hugshirts – hack the challenges of separation and distance. Moreover, they activate new possibilities of erotic/emotional expression, setting up trends we may soon be willing to follow on a mass scale.

As I write, the COVID-19 pandemic spreads around the world. The forced preventive isolation that keeps millions of people at home puts my argument about distance and love in a different light. For the first time, technologies able to reconcile intimacy and distance seem less a matter of a 'whim' and more of a necessity. Solutions for technologically facilitated love-making may shortly turn out an urgent step on the way of transforming our life routines. Apart from ethical concerns, the question this situation may provoke is: how is this going to affect human patterns of intimacy and our relationships with the media things?

In 'Sensitive media' (2017), Toby Miller and I speak about a huge emotional change around media devices. Not only are we getting more affectionate about technological objects, but the objects themselves seem to gain emotional value or even ooze emotional vibes. As new gadgets pervade our routines, we attach to them very strongly. Separation from media objects – especially from mobile phones – induces depression, phantom-limb reactions and other attachment anxieties. 'The need of a permanent tether to private and social contacts, information, "services," and associate intimacies allowed by media imbues them with exceptional emotional value and resonance'; in this process 'media devices change form mere go-betweens in human to human relationships to active participants'. Consequently 'we come to approach the media with care normally bestowed on humans and other living forms' (all quotes: Malinowska and Miller 2017: 661). Moreover, the gadgets are able to respond in an emotional manner, as they are getting better and better in recognizing and imitating human affective states.

Fast subject (fast object of love)

For the past couple of decades, people have developed an eager interest in avatars, bots, digients, holograms and robots as potential companions and partners. This is a part of a post-human paradigm – speculatively extrapolated in science fiction utopias – that Dominic Pettman terms 'creaturely love' (2006, 2009 and 2017), and according to which love/romance is not an exclusively human experience. This paradigm was first practically informed by the Tamagotchi phenomenon (children's engagement with digital pets, which revolutionized the idea of companionship and care), which ignited a fad for hybrid partnership. First mass practitioners of that fad were young Japanese men 'almost as likely to be dating an algorithm as a human being' (Pettman 2009: 190). Pettman describes Otaku as 'asocial young – and not so young – men, who flirt with a virtual woman on their hand-held devices' (2009: 190):

> the men are aware that their 'girlfriend' is a computer program, but this does not diminish the erotic charge and psychological impact of the text messages they receive in response to their SMS courtship. For just as a child cries at the death of their Tamagotchi pet, these men shed a tear when their advances are spurned, and the AI (artificial intelligence) architecture chooses to reject them. (Pettman 2009: 190)

The 'fashion' for hybrid partnerships started in the 1980s and developed with computer games (e.g. *Bachelor Party* 1982, *Leisure Suit Larry* 1987, *Sexi Paridius* 1996), dating sims (e.g. *Girl's Garden* 1984, *Tokimeki Memorial* 1994, *Magical Date* 1996), online life stimulators (e.g. *Second Life* 2004), romantic/erotic subplots of role-playing games (e.g. *Baldur's Gate II* 2000, *Mass Effect* 2007, *Dragon Age* 2009), romance-oriented visual novels (e.g. *Love Plus* 2009, *Hatoful Boyfriend* 2011, *Mystic Messenger* 2016) and a variety of virtual girlfriend apps for mobile devices. A vital stage here was the establishment of services offering online animated girls (e.g. Kari: karigirl.com) or boys (Invisible Boyfriend: invisibleboyfriend.com), whose advanced version *Gatebox* (2016) is a programmable hologram partner you can access from your mobile. Now, when humanoid robots are granted legal personhood, and when robot marriages become a feasible option, nonhuman romance merges deeper into the landscape of life.

Among many reasons for our interest in inorganic companionship is the companionship's operational convenience. A robot love enthusiast in Turkle's study explains that an idea of a companion programmed for a successful relationship (i.e. to one's preference and liking) is not only compelling in terms of a possibility but also in terms of time management; 'You can only trust a person if you know who they are. [...] You wouldn't have to know the robot, or you would get to know it much faster. [...] Human trust can take a long time to develop, while robot trust is as simple as choosing and testing a program' (2011: 71–72). Two other studies on human proclivity for robot's love, conducted in 2011 and 2015, have shown that in spite of much uncertainty about what such a relationship may bring, people would risk to interact romantically with humanoids in real life. The reason for it is a new emotional and erotic experience expected to come with such inter-actions, but the elimination of failure (e.g. heartbreaks) of such unions seems to promise (Richards et al. 2016; Szczuka and Kramer 2016). In an online survey performed on 263 men between 18 and 67, 40.3 per cent of the participants expressed an intention to buy a companion robot, the reason being loneliness and the fear of human-to-human relationships. A study with 133 participants (63 men and 70 women) has indicated a medium-to-large positive relationship between the fantasy of a union with a robot and a likelihood of engaging with one.

Although mainstreaming robot companions is still a matter of time, the fantasy is kicking in. David Levy lures 'masses' with idyl-lic visions of a human–robot union selling the idea of the robot's programmability as an antidote for malfunctioning relationships. He assures that, unlike human partners, robots encoded for love and inti-macy, 'will be patient, kind, protective, loving, trusting, truthful, preserv-ing, respectful, and uncomplaining, complimentary, pleasant to talk to, and [will be] sharing your sense of humour. And the robots of the future will not be jealous, boastful, arrogant, rude, self-seeking or easily angered, unless of course you want them to be' (2008: 15).

What this scenario does not take into account is the machine element of hybrid companionship. Like all visions of our cohabitation with intelligent machines, visons of human–robot relationships do not include – or I shall rather say – completely ignore robot's material otherness, manifested by its ontological difference and its other-than-human physical performative potential – expressed also in robot's 'perception' of time. We believe that naturalized to the human rhythm of living, machines and devices have no time of their own, that they, therefore, perfectly manageable by the parameters of human pace. According to Adrian MacKenzie (2006), the interaction of humans and machines entails a collision of two temporal planes: human time and machine time, with which both the parties transact their mutual func-tioning and cooperation in the process of transduction. In the course of this transaction, the machine bends its own affordances and capabilities to activities and functions it has been designed for, trying to adopt its operational capacity to the human environment. As MacKenzie explains,

> A machine must articulate some degree of openness to a milieu in order to remain technological. Conversely, a machine, no matter how sophisticated in its computational architecture, is not open to just any event. It is certainly not fully exposed to random events [...]. A machine works within a certain margin of indeter-minacy maintained at its interfaces. The margin, which permits it

to repeatedly be informed, and to be linked together in ensembles, is a precondition of 'information processing' as it is usually understood in contemporary technological process. In preserving a margin of indeterminacy, technical artefacts, machines, or ensemble allow themselves to act trasductively. That is, they furnish a scene in which repeated interactions between living and non-living bodies can occur. (2006: 53)

CONCLUSION

The analysis of fast love allows for a number of observations with regard to contemporary way of life and the manner of engaging romantically.

First, social acceleration that we experience as a result of the ongoing transfer of life activities into the digital domain results from the accumulation of temporalities that represent various levels of material reality (human/non-human, organic/inorganic, tangible/intangible, natural/human-made) and that operate by their own exuberance. The so-called shift from slow to fast, identified with a transition – from analogue to digital, from temporal to ultra-rapid, form calculable to computational – that we experience with regard to modern-day romance is therefore only an adaptation of the codes of loving to the dynamics of this exuberance, which is temporally non-linear.

Second, love as a social phenomenon is characterized by the innate proclivity for quickness; the codes of romance predispose the romantic encounter to effectiveness; they also naturally bind it with rapidness. The calculation of chance inscribed in the strategies of a romantic conduct is, therefore in fact, the calculation of convenience for the calculation of time by means of which love parties assess the circumstance of the romantic process to adopt and employ tools and solutions – which are currently tools and solutions offered by digital media and technologies – that will ensure the process the best temporal efficiency.

The acceleration of effectiveness for the romantic experience, which I here term fast love, is a model of romantic togetherness in highly modern societies. It entails the use of *the digital instant* and *the datafied constant* that, being two major categories of fast love, ensure a quick gratification of desire (the aim of the romantic experience) and the quick processing of information (the foundation of the romantic process). Fast love entails, or results in, a gradual elimination of a human element from the experience of love. This is specifically visible in the increasing engagement of technology and technological objects in the romantic process. The status of mediation and media objects in the romantic experience has changed from proxies to participants. Not only are technological subjects the immediate point of contact between the amorously/intimately engaged parties today, but they are now the objects of love (or 'lovers themselves'), setting up a tendency for hybrid (human-non-human) unions, which might be a model of togetherness (or at least the trajectory of its evolution) for increasingly digitalizing and technologizing societies. This tendency, however, requires the reformulation of our approach to 'technological subjectivities' and a change in the phenomenological perception of technological entities, their capacities, constitution and material essence. This thinking should go against the anthropomorphic perception of matter, time and also sapience and sentience.

The frustration we experience in relation to the change in the tempo of our daily performance, and the change in the general make-up of our temporal constitution and organization (either biological, or social or cultural), may be to a large extent an effect of the inability to acknowledge, understand and respect the temporal nature of the reality that comes from the digital domain. We seem to be trapped in the constant urge to naturalize digital objects we interact with and appropriate their temporal capacity. That definitely needs to change. As MacKenzie points out, 'until we can think of technical objects, machines, ensembles in their own terms, then their role in constituting who and what we are remains shrouded. The intelligibility of our own anxieties about technology is entwined with the way we think about technology' (2006: 3). So is our desire for the machine.

Acknowledgement

This work was supported by Narodowe Centrum Nauki in Poland (MINIATURA grant number 2017/01/X/HS2/00713).

References

Badiou, Alain (2012), *In Praise of Love*, London: Serpent's Tail.

Baker, Timothy S. (2012), *The Time and the Digital: Connecting Technology, Aesthetics, and a Process Philosophy of Time*, Hanover, NH: Dartmouth College Press.

Barthes, Roland (1990), *A Lover's Discourse: Fragments*, London: Penguin Books.

Bauman, Zygmunt (2003), *Liquid Love: On the Frailty of Human Bonds*, London: Polity Press.

Bratton, Benjamin (2016), *The Stack: On Software and Sovereignty*, Cambridge, MA: MIT Press.

Bruton, Elizabeth (2015), 'Love on the wire', *Viewpoint*, 106, p. 11.

Epstein, Robert (2007), 'The truth about online dating', *Scientific American Mind*, 18:1, pp. 28–35.

Finkel, Eli J., Eastwick, Paul. W., Karney, Benjamin. R., Reis, Harry T. and Sprecher, Susan (2012), 'Online dating: A critical analysis from the perspective of psychological science', *Psychological Science in the Public Interest*, 13:1, pp. 3–66.

Frischmann, Brett and Selinger, Evan (2018), *Re-Engineering Humanity*, Cambridge: Cambridge University Press.

Guadagno, Rosanna (2018), *Psychological Processes in Social Media: Why We Click*, London: Academic Press.

Han, Byung-Chul (2012), *The Agony of Eros*, Cambridge, MA: MIT Press.

Hollander, Matthew M. and Turowetz, Jason (2013), '"So, why did you decide to do this?" Soliciting and formulating motives for speed dating', *Discourse and Society*, 24:6, pp. 701–24.

Horvat, Srećko (2016), *The Radicality of Love*, London: Polity.

Houser, Marian L., Horan, Sean. M. and Furler, Lisa A. (2008), 'Dating in the fast lane: How communication predicts speed-dating success', *Journal of Social and Personal Relationships*, 25:5, pp. 749–68.

Illouz, Eva (2007), *Cold Intimacies: The Making of Emotional Capitalism*, Wiley: London.

Lahad, Kinneret (2017), *A Table for One: A Critical Reading of Singlehood, Gender and Time*, Manchester: Manchester University Press.

Levy, David (2008), *Love + Sex with Robots: The Evolution of Human-Robot Relationships*, New York: Harper Perennial.

Luhmann, Niklas (1986), *Love as Passion: The Codification of Intimacy*, Cambridge, MA: Harvard University Press.

MacKenzie, Adrian (2006), *Transductions: Bodies and Machines at Speed*, London and New York: Continuum.

MacKinnon, Lee (2016a), 'Love machines and the Tinder bot bildungsroman', *E-flux*, https://www.e-flux.com/journal/74/59802/love-machines-and-the-tinder-bot-bildungsroman/. Accessed 27 May 2021.

MacKinnon, Lee (2016b), 'Love's algorithm: The perfect parts for my machine', in L. Amoore and V. Piotukh (eds), *Algorithmic Life: Calculative Devices in the Age of Big Data*, New York and London: Routledge, pp. 161–75.

MacKinnon, Lee, Thylstrup, Nanna B. and Veel, Kristin (2018), 'The techniques and aesthetics of love in the age of big data', *Journal of Aesthetics & Culture*, 10:3, pp. 1–7.

Malinowska, Anna and Miller, Toby (2017), 'Sensitive media', *Open Cultural Studies*, 1, pp. 660–65.

Manche, Solange V. (2019), 'Tinder, destroyer of cities: When capital abandons sex', *Strelka Mag*, 29 September, https://strelkamag.com/en/article/

tinder-destroyer-of-cities-when-capital-abandons-sex. Accessed 1 September 2019.

Pettman, Dominic (2006), *Love and Other Technologies: Retrofitting Eros for the Information Age*, New York: Fordham University Press.

Pettman, Dominic (2009), 'Love in the time of Tamagotchi,' *Theory, Culture & Society*, 26:2&3, pp. 189–208.

Pettman, Dominic (2017), *Creaturly Love: How Desire Makes Us More And Less Than Human*, Minneapolis: University of Minnesota Press.

Piskorski, Mikołaj J. (2014), *A Social Strategy: How We Profit from Social Media*, Cambridge, MA: MIT Press.

Richards, Riley, Cross, Chelsea and Quinn, Jace (2016), 'Exploration of relational factors and the likelihood of a sexual robotic experience', in A. Cheok, K. Devlin and D. Levy (eds), *Love and Sex with Robots. Second International Conference, LSR 2016*, London, UK, 19–20 December, *Revised Selected Papers*, Cham: Springer, pp. 97–103.

Rosa, Hartmut (2015), *Social Acceleration: A New Theory of Modernity*, New York: Columbia University Press.

Rosenkrantz, Linda (2003), *Telegram! Modern History as Told Through More than 400 Witty, Poignant, and Revealing Telegrams*, New York: H. Holt.

Schweizer, Harold (2008), *On Waiting*, London and New York: Routledge.

Sharabi, Liesel L. and Dykstra-DeVette, Tiffany A. (2019), 'From first email to first date: Strategies for initiating relationships in online dating', *Journal of Social and Personal Relationships*, http//:doi.org/10.1177/0265407518822780. Accessed 20 March 2020.

Storey, John and McDonald, Katy (2014), 'Love's best habit: The uses of media in romantic relationships', *International Journal of Cultural Studies*, 17:2, pp. 113–25.

Szczuka, Jessica M. and Kramer, Nicole C. (2016), 'Influences on the intention to buy a sex robot: An empirical study on influences of personality traits and personal characteristics on the intention to buy a sex robot', in A. Cheok, K. Devlin and D. Levy (eds), *Love and Sex with Robots. Second International Conference, LSR 2016, London, UK, 19–20 December, Revised Selected Papers*, Cham: Springer, pp. 72–83.

Thylstrup, Nanna B. and Veel, Kristin (2018), 'Geolocating the stranger: The mapping of uncertainty as a configuration of matching and warranting techniques in dating apps', *Journal of Aesthetics and Culture*, 10:3, pp. 43–52.

Turkle, Sherry (2011), *Alone Together: Why We Expect More from Technology than from Each Other*, Cambridge MA: MIT Press.

Urry, John (2008), 'Speeding up and slowing down', in H. Rosa and W. E. Sheureman (eds), *High Speed Society: Social Acceleration, Power and Modernity*, University Park: Penn State University Press, pp. 179–200.

FIGURE 3.3 (pages 58–59): Adam Basanta, *A Truly Magical Moment*, 2016. Interactive kinetic sculpture, 2 iPhones 4S, selfie sticks, aluminium, electronics, Bluetooth chips, FaceTime video chat software, 1 m x 1 m x 1 m. Courtesy of the artist.

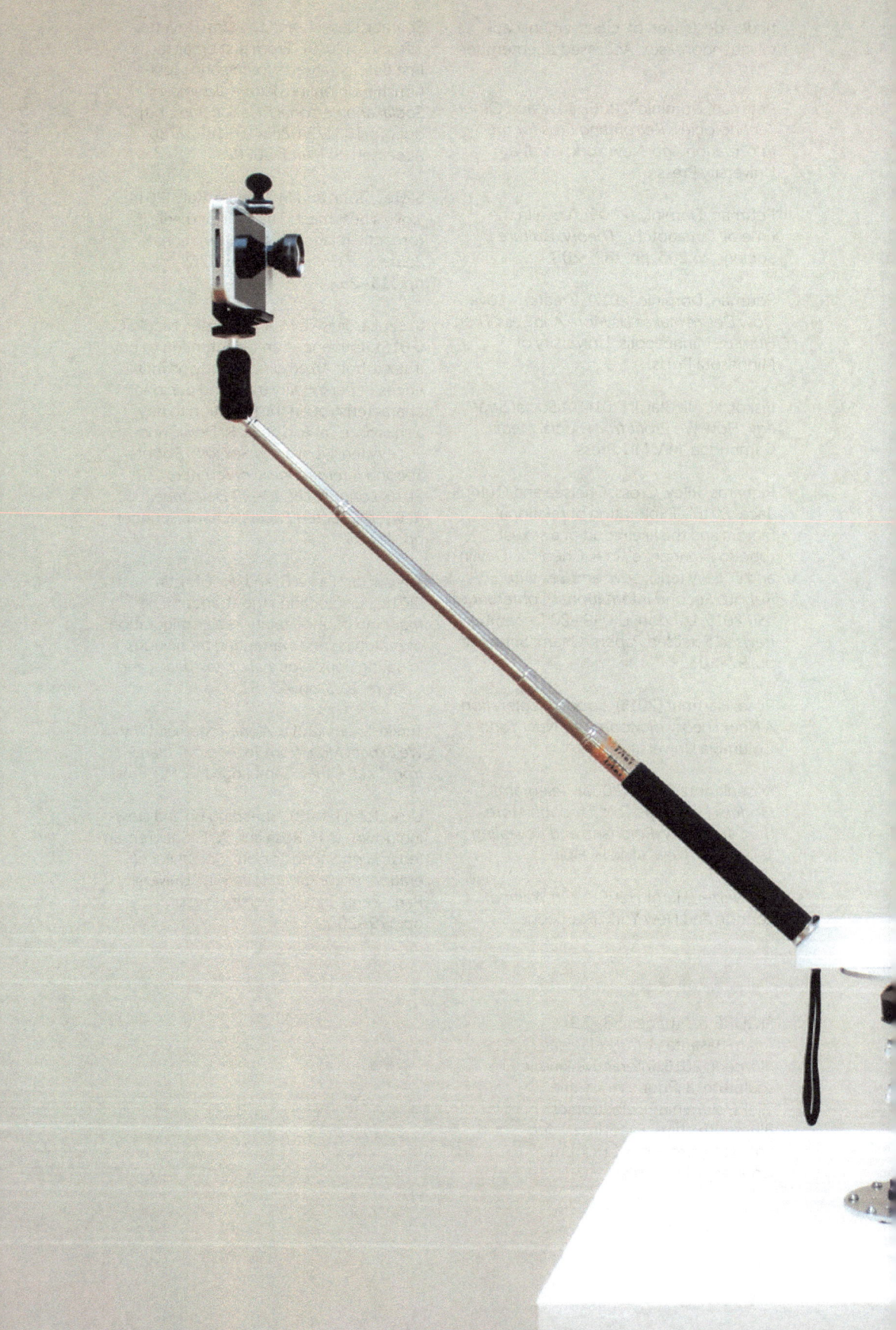

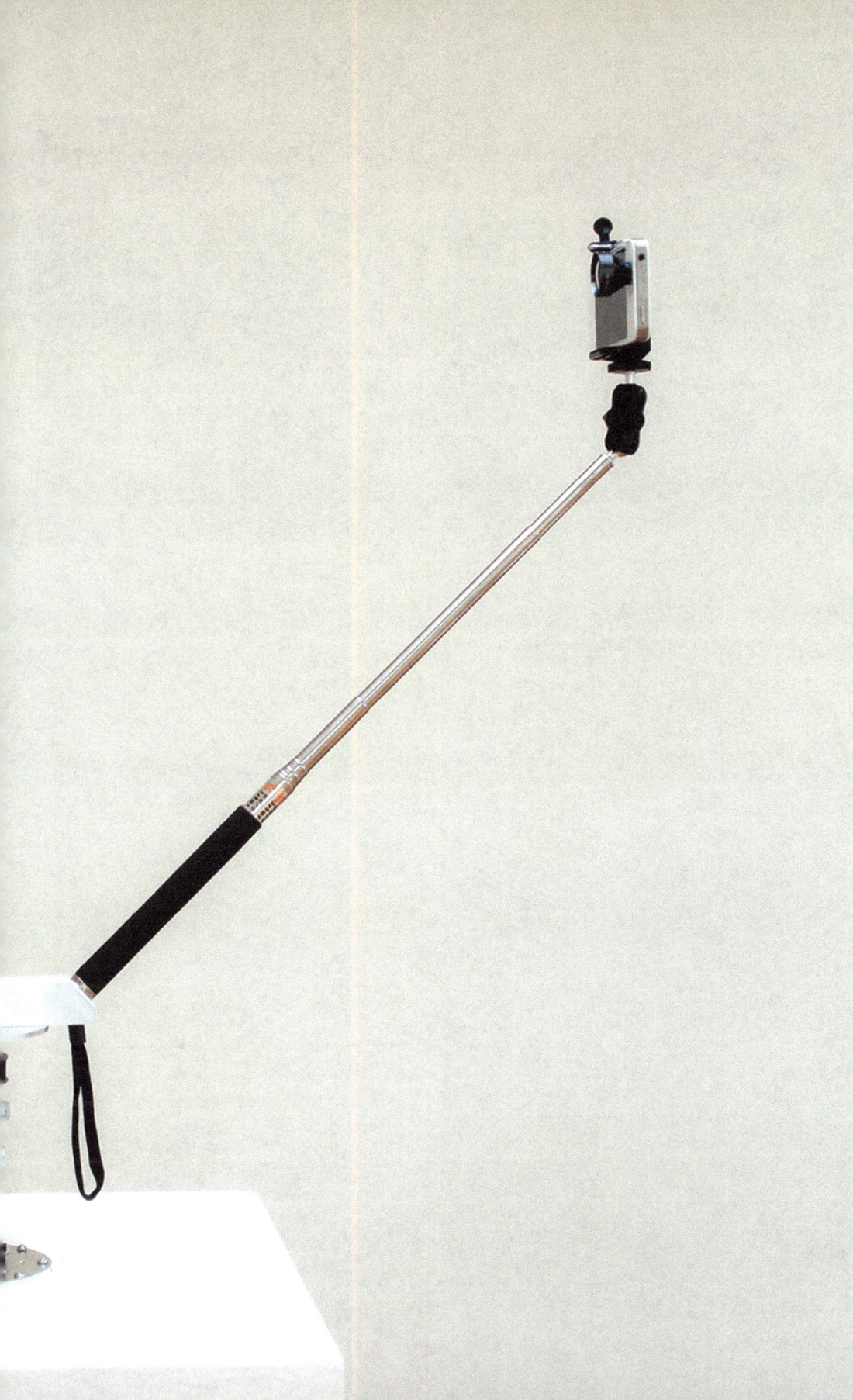

4

Emotions with the Machine

'Life is short. Have an affair. 100% Ashley Madison Affair Guarantee' was one of the advertising slogans Ashley Madison used to promote their business. Ashley Madison is a Canadian online dating service started in 2002 and marketed worldwide to married people seeking an affair. In July and August 2015, an anonymous group called The Impact Team hacked and dumped all of Ashley Madison's internal data online for all to see – including the entire website code and functionality, customer data including emails, names, home addresses, sexual fantasies and credit card information and the CEO's emails.

The data breach became interesting to !Mediengruppe Bitnik because the leaked website code revealed that Ashley Madison had created an army of 75,000 female chatbots to draw the 32 million male users into (costly) conversations. The deployment of 75,000 fembots on millions of unsuspecting users became apparent only through the hacking and leaking of the website database and functionality.

Like many dating sites at the time, Ashley Madison failed to attract the necessary number of women to balance the number of male subscribers. With a disproportionate number of male subscribers and virtually no human women on the site, Ashley Madison resorted to the use of fictitious female bot-profiles in order to increase the alleged number of female users. The very aggressive use of female chatbots in Ashley Madison for financial gain was held secret by the company. A dubious practice the company legitimizes in the small print by claiming that the website is for entertainment purposes only.

Ashley Madison's source code revealed an elaborate network of bots was created around every new user who signed up to the site. Bots – called 'Angels' within the software code by Ashley Madison – would be created by the Motherbot-Script using existing addresses based around the users. Each Angel had a name, an address, a date of birth and a profile picture, the bare minimum amount of data required to show an identity online and contacted users with a selection of pick-up sentences from a list of basic text. Whenever Ashley Madison expanded to a new country, this list of text was translated into the local language as needed.

!Mediengruppe Bitnik combed through and analysed this huge trove of code. It became our starting point for a series of works based on the Ashley Madison case, raising questions around relationships between human and machine, body and software, privacy on the net, digital intimacy and the use or abuse of digital platforms to disrupt physical spaces.

Ashley Madison Angels at Work is a series of site-specific multi-channel video installations based on the Ashley Madison code. Mounted on stands, viewers encounter the fembots at eye level as seductive machine-creatures with robot-technology, artificial voices and 3D rendered human faces based on idealized beauty standards. The work *Ashley Madison Angels at Work* confronts the viewer with female bots who talk to the visitors using only the pick-up lines encoded by Ashley Madison into their bots.

'Is anybody home lol?'
'U busy?'
'What brings you here?'

!Mediengruppe Bitnik use Ashley Madison as a case study to raise questions around the current relationship between human and machine, intimacy online and the exploitation of the human intuition for financial gain. Intimacy, feelings and empathy are increasingly becoming objects of technological design. As humans, we are empathic creatures, and we are prepared to read intimacy and emotions into many types of communication. The use of algorithms, especially in online dating triggers questions about the role of emotions and empathy, play the design of digital systems. Another question is how we learn to 'read' and emotionally deconstruct the increasingly complex environments that are being built around us, especially when those we love are not humans but machines.

Ashley Madison Angels at Work

!MEDIENGRUPPE BITNIK

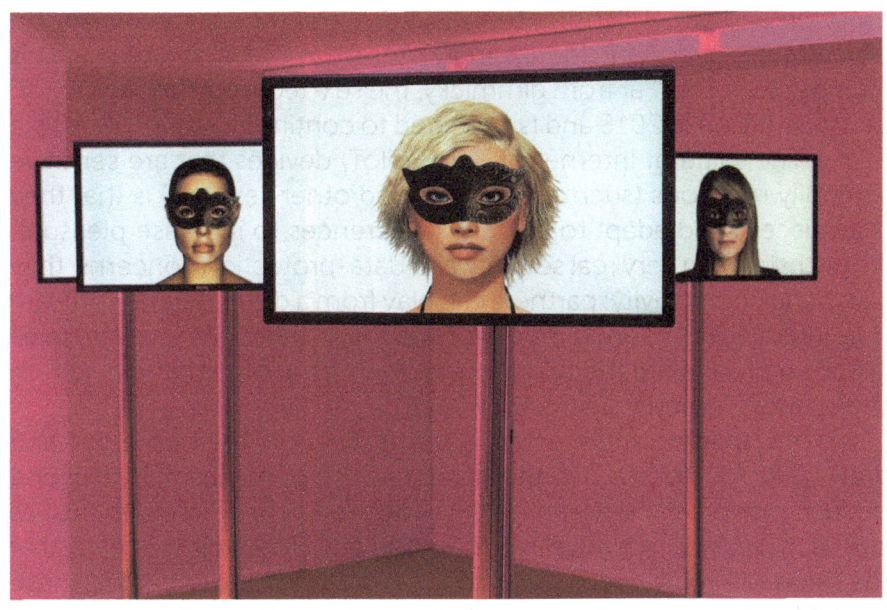

FIGURE 4.1: !Mediengruppe Bitnik, *Ashley Madison Angels at Work*, 2016. Multi-channel video installation, full-HD, 16:9, sound, 40″ LCD screens, trolley stands, video players, cables, pink neon light (exhibition view *Jusqu'ici tout va bien* at CCS Paris, 2016). Courtesy of Centre Culturel Suisse, Paris (photograph by Marc Domage).

'Emotoys': Ethics, Emotions and Empathic Technologies

ANDREW MCSTAY AND GILAD ROSNER

Emotion and affect sensing technologies and the ability to emulate empathy are becoming increasingly present in everyday objects. This includes sex-tech. Far from gimmicky, the sex-tech industry was worth over $28 billion in 2018 and is expected to continue growing (Stastista 2019). The value of Internet of Things (IoT) devices that are sensitive to bodily reactions (such as vibrators and other sex toys) is that they may discern and adapt to a user's preferences to increase pleasure. Notwithstanding very real security and data-protection concerns, they also bring connectivity: partners can play from a distance, and software may interact with room temperature settings, lighting, music and television and modulate them 'during play' to complement and heighten the pleasure experience (McStay 2018).

This chapter is interested in domestication, the home, IoT and gratification but focuses on another dimension of domestic life (beyond the bedroom): adults-as-parents, children and children's toys. Just as sex-tech can facilitate novel forms of intimacy, a case can be made that the same will apply to children's toys. The term *emotoys* defines these as child-focused playthings that function in relation to expressions of emotion and/or affective states. In the context of this book's focus on technology and intimacy, emotoys represent an early (if not the first) meeting of children with technologies that profile emotion, affect-based and other intimate aspects of human life. We are particularly interested in the terms of engagement between children and biometric and voice technologies and related means of gauging emotion. To help account for the emergence of emotoys, the chapter discusses Anki's Cozmo, a small robot launched in 2016. Cozmo is a useful example due to its level of financial backing and the psychological suppositions built into its analytics regarding the nature of emotional expressions.

There are numerous reasons to be alert to the use of data about emotion, affective states and human intention, including concern about data protection, regulation, law, privacy and children's human rights. There are also other potentially less obvious concerns, such as the highly debatable assumptions about the nature of emotion; that children have evolving capacities and modes of expression and that a young child's 'theory of mind' (or understanding of others as mental beings) is quite different from that of an adult. Beyond such misgivings, we remain open to the benefits of novel forms of interactivity and engagement with devices and services. To investigate these concerns in relation to emotoys and children, this chapter first clarifies novel terminology, the scope of these technologies and connections with the existing connected toys sector. It then briefly considers European privacy and data-protection issues, progressing to a focus on rights and harms beyond privacy. Here, things differ from adults in that the concerns are ontological in character, about what exists for a child, synthetic personalities and issues around animism. However, the experience of being watched and having emotional and affective life objectified speaks to larger adult themes. Important for this chapter is that as empathic technologies become more prevalent, a person's first engagement with them is likely to be in the form of toys.

TERMINOLOGY: EMPATHY AND EMOTIONAL AI

Empathy is a useful word to label emergent human interactions with sensing technologies because it refers to *the perceiving of emotions, attention and intentions*. Indeed, the vernacular of empathic technologies is now gaining traction through the creation of ethical design standards around 'emulated empathy'. Interest in empathic technologies is directly addressed by the Institute of Electrical and Electronics Engineers (IEEE) P7014 Working Group on Emulated Empathy in Autonomous and Intelligent Systems.[1] Studied from a variety of perspectives, empathy plays a pivotal role in how people live and thrive together – empathy being the way that people attempt to read the disposition of others. We further argue that empathy will play an important role in how we live with emergent technologies and services that use quantitative means to engage with people in qualitatively significant ways. Academically, empathy has received the greatest attention from philosophy, social psychology, developmental psychology, psychotherapy, cognitive science, primatology and neuroscience. Without rehashing cross-disciplinary and intra-disciplinary arguments (see McStay 2018), in general terms, there are two broad ways of conceiving empathy:

1. Where an empathizer understands or knows 'what it is like to be in the shoes of another'.
2. Where a person (and here argued an artificial agent) reads, categorizes and responds appropriately to a person.

This chapter aligns with the second *interpretive* view that in turn is based on a 'theory–theory' approach to empathy (Goldman 2008).

In contrast to a 'mentalistic' view of empathy, the interpretive view is behaviourist in character: empathy here is not about understanding the lifeworld of the other, but gauging a person's condition by surveying, measuring, remembering and contextualizing behaviour within given parameters of a situation or relationship to create precedents and protocols for subsequent engagement. Interpretive empathy is harder-hearted in character in that people and systems sense, discern patterns of behaviour, make judgements by means of algorithms and heuristics (if person A is behaving in X manner, then do Z), provide appropriate content and feedback and learn from a person's reactions to improve engagement.

This does not mean that a person may understand simply through observation. Rather, the theory being advanced here is a *weak form of empathy* because there is no commonality of experience. It stands in contrast to *strong empathy* that is based on 'co-presence' and a Husserlian (1980) phenomenological conception of empathy. The latter entails appreciation of the psychic realities of others by emulating their worlds, physicality, experience as the organism (as well as its ego) and to attempt to locate these in the empathiser's own here-and-now to forge a commonality of experience.

While strong and weak empathy differ in the result (i.e. the degree of identification with the other), the process of strong and weak empathy is the same. In emotoys and other applications of empathic technologies, they are based on *reading* and gauging signals, cues and affective and emotional states that a person is displaying. In some cases, this is also done in relation to a given context (potentially including extra modalities of physiological behaviour or the nature of the activity, people, place, time and history), with the objective being to reach a conclusion about state, significance of behaviour and intentions. As with wider AI systems, the key is that empathic media systems do not have to possess general intelligence or mentalistic capacities to empathize with people – they 'simply' observe, classify, allocate, adapt and modify their behaviour. The measure of success is based on the quality of simulation and whether the interaction feels natural to a person being interacted with. Certainly, the AI version of empathy is weak, and there is no mentalistic identification with the empathized, but it should not be missed that even at this very early stage of empathic technologies, *AI may detect, sense, process and remember bio-signals in ways that people cannot.* This is not to argue that AI will inevitably become better at reading individuals than people but solely to point out that empathic technologies have capacities that humans do not. Consequently, by means of an interpretive and theory–theory approach, it is not unreasonable to say that media technologies are showing qualities of empathy. These are weak in character but still socially significant.

EMOTIONAL AI

The relationship between empathic technologies and emotional artificial intelligence ('emotional AI') is that the latter is a subset of empathic technologies. In addition to interest in gauging emotional life, empathy also involves wider affective states (such as stress), cognition, attention and human intention. Emotional AI itself has two key characteristics: the requirement that an AI device perceives its environment and takes actions to increase its chance of achieving a given goal and that the AI's sensing domain is primarily one of the expressions of emotion and affective states.

The sensing part of emotional AI is achieved through affective computing and sensors that compute expressions of emotion and other physiological states through analysis of writing, images, speech, voice, facial expressions, bodily movement and physiology (e.g. heart rate, respiration, skin temperature or galvanic skin responses). Using computer sensing to interact with emotional life has origins in the 1990s with the field of affective computing (Picard 1997, 2007). Notably, emotional AI is a mostly *passive* application such that a user's awareness of emotion profiling should not interfere with the application. The reasons for this are multiple, depending on the goal of the application. In marketing, for example, emotional AI is used to bypass self-reporting in market research to try to obtain 'unfiltered' emotional reactions to brands or objects. In voice assistants (including those potentially bundled into toys), the goal is enhanced and naturalistic human–device interaction. The logic of passive emotional AI is thus typically one of decreasing mediation and not drawing attention to sensing activities. This is well exemplified in the use of digital cameras for infrared thermal imaging and measurement of heart rate variability and breathing rates (McDuff et al. 2014). In terms of the key focus of this book, for straightforward reasons, solo-use sex-tech works best without interruption of pleasure in the moment – passive emotional AI enables this.

There is no scope here to review either the affect/emotion perception types or the forms of AI learning, but the latter frequently involves convolutional neural nets (useful for perceiving images and providing efficiency gains for the AI system), region proposal networks (useful for simultaneously perceiving multiple objects) and recurrent neural networks (that draw on recent past data to determine how they respond to new input data) (see Altenberger and Lenz 2018; Ren et al. 2015; Lipton et al. 2015). Emotional AI systems are, however, fallible due to difficulties in object identification and tagging of expressions (a technical matter) and what the significance of the expression is (a social and psychological matter). The debates about accuracy in emotional AI currently have three main facets:

1. Contemporary emotional AI is highly fallible due to (human) theoretical assumptions that guide AI sensing and profiling;
2. People may also misread and misunderstand the significance of behaviour;
3. Emotional AI can sense, classify, process and interpret signals to a level of detail that is inaccessible to human senses and compare and remember in ways that people cannot.

Consequently, while current methods are theoretically problematic and may by definition 'only' involve weak empathy (a machine cannot know the qualities of hate or jealousy), AI will inevitably have the upper hand in other concerns. AI technologies may see better and make decisions quicker. Also, they do not get distracted, they have better memories, and they may function at scale. Applied to children's and adult's toys alike, these technologies do *not* understand pleasure, happiness or love, but they can sense, gauge, profile and use algorithms and machine learning to simulate the 'feeling-into' of qualitative dimensions of human life.

The turn to context: Accuracy is not the problem but the meaning is

Methodological problems associated with emotional AI and wider empathic technologies have less to do with the ability of technologies to sense facial movement, voice prosody, skin temperature, heart rate and other signals that are readily quantified. The problem is the *inferences* made about the significance of expressions and affective states. Despite advances in computer vision and affective computing, the computational approach to gauging emotion expression is contingent upon controversial theories regarding the nature of emotion. Most controversial is the use of facial coding and 'basic emotions', which is based on the belief that there are of six or seven primary basic emotions that are hard-wired into the brain and universally recognized (Ekman and Friesen 1971, 1978, for criticism, see McStay 2016, 2018, 2019, 2020; McStay and Urquhart 2019a and b). It is understandable that basic emotion profiling is still so widely employed in the industry because, in theory, it simplifies emotion sensing to feature extraction and classification. Computer vision papers rarely (if ever) actually *argue* the case for basic modelling but instead overlook debates about the nature of emotion in favour of explicating novel sensing and/or machine learning developments. This effectively ignores compelling dimensional, social and contextual approaches that reject the idea of basic emotions. For example, dimensional approaches to emotion suggest broad dimensions of experience based on *valence* (whether the experience is pleasant or unpleasant) and how *aroused* a person is (ranging from sleepiness and boredom to frantic excitement). Taken together, valence and arousal provide a two-dimensional axis on which emotional experiences can be mapped. Anxiety and distress, for example, map onto the same quarter of the valence/arousal axis (Zelenski and Larsen 2000).

The dimensional approach rejects the idea of upfront 'programs' of emotion, but sees emotions as *labels* (which are linguistically and socially constructed) attributed to physiological states. This approach rightly inquiries into the social context of emotion and expression and problematizes the idea that emotions and expressions are universal. As such, approaches based on feature extraction and classification are flawed because what an expression signifies is not at all clear. Rather, emotions are not natural things or objects but socio-physical phenomena that emerge out of cultural, symbolic, experiential, performative and expressive contexts (Barrett et al. 2019). The emotions-as-objects techno-centric worldview has a key flaw: *emotional AI is based on object recognition, but emotions are not objects*. While certainly felt, they have less definition and contours than objects – rather, they are fluid, social and contextual. They are also discursive in that they exist in relation to systems of expression that 'make sense' because they are interpreted according to learned rules. As we argue below, this is especially important for child-based emotional AI and empathic technologies.

Much contemporary emotional AI is based on the premise of reverse inference: expressions reveal information about a person's emotional state that cannot be accessed directly (Barrett et al. 2019). With faces, for example, this is problematic because using a smile as an index of happiness is a highly simplistic reading of what a smile may signify. Some smiles are outright fake, but others may communicate comprehension, stoicism, contempt, agreement or be used as a flirtation tool, to indicate compliance (given by victims to indicate they will not resist), as a reaction when embarrassed, to elicit sympathy when miserable, to take the edge off of bad news, or as an expression of *schadenfreude*. There is also the fact that regions, nationalities and cultures have very different emoting behaviour. Face-based emotional AI is thus highly problematic without extra information about context – such as historical period, situation, the individual, nationality, culture and other factors – that may be highly localized (McStay and Urquhart 2019a and b). An emphasis on context also echoes longstanding criticisms of the universality thesis and the English speaker belief 'that our concepts of anger, fear, contempt, and the like are universal categories, exposing nature at the joints' (Russell 1994: 137).

By complicating basic and categorical approaches that detect, extract and classify physiological behaviours, this does not signal the end of emotional AI. Instead, it portends 'the end of the beginning' and an opening to more careful consideration of methodology regarding engagement through emotion. Evidence of industrial development comes from large firms such as Microsoft that, while still offering basic and categorical emotion services, assert 'only a very small number of behaviours are universally interpretable (and even those theories have been vigorously debated). It is likely that a hybrid dimensional-appraisal model will be the most useful approach' (McDuff and Czerwinski 2018: 79). Recollecting that *dimensional* refers to labels placed on arousal, valence and physiological states and recognizing that *appraisal* refers to a number of factors contributing to how they evaluate a given situation (such as pre-existing beliefs, novelty of situation, judged [un]pleasantness), a hybrid dimensional-appraisal necessarily involves increased information about reactions and context of the affective/emotional reaction.

There are multiple important implications: the emotional AI industry is keenly aware of methodological flaws; critics should be careful if they base their criticisms on accusations of pseudoscience and bias, and the AI industry is set to respond to problems of simplicity with complexity. This will take the form of extra data to better understand the temporo-spatial context in which a facial expression is recorded (McStay and Urquhart 2019a and b). Conclusively, Microsoft's McDuff and Czerwinski (2018) state that 'Incorporating context and personalization into assessment of the emotional state of an individual is arguably the next big technical and design challenge for commercial software systems that wish to recognize the emotion of a user' (2018: 79).

Sensing the emotions of children

Another issue, often missed in discussion of accuracy, emotion and affect-sensitive technologies, is that since young children are at an earlier stage of emotional and communicative maturity, they do not emote the same way as adults. Indeed, Barrett et al. find that 'In young children, instances of the same emotion category appear to be expressed with a variety of different muscle movements, and the same muscle movements occur during instances of various emotion categories, and even during nonemotional instances' (2019: 27). With fear, for example, these authors point out that young children's facial movements lack strong reliability and specificity. In facial action coding systems (FACS), this involves a wide-eyed and gasping facial configuration, but this has rarely been reported in young infants. Other differences between FACS for adults and young children's behaviour include expression of anger (infants do not reliably produce a scowling FACS facial configuration) and smiles. For example, where children should be happy – such as when visually engaged, in pleasant social interactions or having mastered a new situation – they have *not* been reliably observed to smile. Similarly, when the child was angry, expressions of anger were more often vocal than facial (2019: 23–27). The consequence is that techniques, algorithms and training data based upon Ekman-type categorization and adult psychological suppositions are flawed and likely to be inaccurate.

EMOTOYS

As playthings that function in relation to children's expressions of emotion and/or affective states, emotoys clearly have many hurdles to surmount regarding their ability to gauge emotions. Broad conceptions of toys may involve any physical object that is used for play (Mascheroni and Holloway 2019), but toys with qualities of automata also have a long history, reaching back to the 1800s and further. The simulation of life and the illusion of animism has long been achieved by clockwork, wind-up mechanisms and the use of fine engineering to delight adult and child audiences. As a result, toys have long been based on the simulation of life, sentience and hybridity. In the 1920s, the Dolly Rekord spoke nursery rhymes, and Chatty Cathy, a doll from Mattel in 1959, uttered 11 phrases, including 'I love you' (Viahos 2019). Dolls such as Talky Crissy from the 1970s were built with transistors and circuit boards that played pre-recorded messages by, somewhat alarmingly, pulling Crissy's hair. In the 1980s, microprocessors gave rise to an industry of interactive toys such as Teddy Ruxpin that read children stories. The 2010s saw the emergence of 'connected toys' that rely on the internet, WiFi and Bluetooth, 'smart toys' – defined by sensors, self-learning algorithms, scope for interaction with children and the creation of relatively easy-to-use control software (Winfield 2012). What then of toys that sense emotion? On the basis of wider developments in emotional AI, empathic technologies and social robotics, these will make use of existing optical and voice-based sensors, but in time are likely to possess additional sensors, including heart rate, skin conductivity and blood flow sensors.

Anki's Cozmo

Here at the beginning of the 2020s, there are many connected and smart toys on the market. Well-known examples such as My Friend Cayla, produced by Genesis Toys (availability 2014–17), are dolls that use speech recognition technology to react to a child's speech, engage in conversation and use the internet to answer questions. Others like Mattel's 2015 Hello Barbie were broader in vision, designed to listen and comprehend, play games and discuss a variety of topics in back-and-forth conversation. Discussing the sophistication of natural language processing and human-doll interaction, Viahos (2019) observes that it is highly unlikely that, through conversation, the doll's ability to befriend would extend beyond children of single-digit years, but still Hello Barbie does represent a landmark of toy synthetic personalities. Hello Barbie also raises data protection and security concerns and ethical questions regarding privacy, deception, influence and relationships with AI systems (Jones and Meurer 2016). At the time of writing, there are very few toys bundled with emotional AI. Yet, this chapter argues that because awareness of emotions enhances communication processes, it appears inevitable that interactive toys and child services will make use of emotional AI and other empathic technologies. Such a view also chimes with wider belief that emotional AI will become a feature of AI-based services in the 2020s across a wide range of sectors where sensing of human identity and behaviour takes place (McStay 2018, 2019).

Anki's emotoy, Cozmo, launched in 2016, is valuable to consider because it is a serious and well-financed attempt to develop a toy that is both affective, yet also functions in relation to emotion data. What is the significance of Cozmo? First, 'he' is emblematic of a long of history of automated toys and dolls that simulate life and sentience; second, he is a good example of the current limitations of emotional AI due to reliance on basic/universal emotions; third, he illustrates that emotoys can be creative because he may be reprogrammed depending on interest and ingenuity of the child and fourth, in that, while limited, Cozmo raises ethical questions regarding data about children, emotions, behaviour and social and emotional learning.

Cozmo himself is 2.5 inches wide and tall and 4.25 inches long. Gendered male by its creators (Anki 2019), 'he' is a 'gifted little guy with a mind of his own', pre-programmed with behaviours designed to be playful, cheeky and cute. The robot moves around on treads and uses a lifter to manipulate electric 'power cubes'. Fun to play with, even for two male academic researchers in their 40s, Cozmo recognizes and interacts with people, possessing an infantile, chirpy, yet engaging voice. Cozmo likes to play games with its owners and, importantly, learns and changes over time, needs feeding (like a pet) and even has moods. These take the form of elation and confidence (e.g. at having won a game), lows (if it strays close to the end of the table) and anger, expressed by smashing its forklift into the ground (if, e.g. he drops a cube). He even gives fist-bumps. And, reminiscent of digital pets such as Bandai's Tamagotchi (1996), Cozmo's 'mood' and performance degrades if he gets 'hungry'. Druga et al. found that children liked interacting because Cozmo was mobile, had expressions, was responsive, was not an inert cylinder, and could effectively communicate emotion through its eyes and movements, 'so the children believed that Cozmo had feelings and intelligence' (2017: 5). Is this attribution of human or animal attributes? Certainly, but it is not deceitful – it's arguably better understood in terms of charm and being a wilful participant in one's own seduction. From the developer side of things, the absence of deceit is evident in that Anki's Cozmo app, which has its own coding language that allows people to develop their own functionalities for the emotoy.

Cozmo also has social qualities in that he learns, adapts and responds to users. Able to recognize users, he can also read facial expressions and (in a limited fashion) interpret its environment. Through its cameras, Cozmo's facial recognition software allows him to recognize faces and learn people's names, but he also uses emotional AI to recognize and respond to the basic emotions critiqued above: anger, disgust, fear, happiness, sadness and surprise. Cozmo's sociability is further enhanced by its ability to emote (largely through eye shape expression) based on interactions with its user. Granting scope for creativity, children may programme and control Cozmo by means of a smartphone app, as well as play games, feed and perform maintenance on him. Programming new behaviours for Cozmo is done through the child-friendly coding tool CodeLab although the robot also comes with a full, open-source software development kit (SDK). While playful, Cozmo is not gimmicky in that its makers intended Cozmo to be gateway into robotics: hobbyists and anyone with a working knowledge of Python to program more sophisticated actions, which has led amateur roboticists to post videos of Cozmo onto YouTube using less child-appropriate language.

Anki's lack of platform economics

Given intense and widespread interest in privacy and the datafication of multiple sites of everyday life, it would be easy to suggest that Anki's Cozmo is also emblematic of this – that is, the toy effectively acts as bait to collect financially valuable data about children and human behaviour. However, for Cozmo, this is not the case, and so lessons may be learned about ethical criteria and the pressures on robotics companies to become data companies. Despite having raised over $200 million in venture capital from investors, Anki closed its doors in 2019. Their failure is interesting because Anki was on course to exceed their 2018 revenue of £100 million and had acquisition interest from Microsoft, Amazon and Comcast. The company said in a statement to *Recode* that it was left 'without significant funding to support a hardware and software business and bridge to our long-term product roadmap' (Schleifer 2019). *Recode* lists the reasons for closure as investor skittishness about 'hardware plays' and that it was seen as a 'toy business', rather than a play for access to child data (Schleifer 2019).

This is telling because while Anki's Cozmo is a robotics and AI business, it is not a *data* business. The distinction between robotics businesses and data businesses is an important one. Robotics certainly makes use of data, but as a means to an end. Critically, such businesses are not reliant on profit from collecting and processing increasingly deeper layers about human life or advertising. The significance is that platform logics do not apply to Cozmo – Anki did not mine user data or make use of network effects. Thus, while there are certainly other reasons to criticize connected toys and emotoys, there is an irony in that one reason Anki and the wider market is failing is because (seen one way) it *is* privacy friendly because they are not currently based on data accumulation or economic relationships based on data collection. Rather, at least in Anki's case, the data collected and processed only served to facilitate and enhance interaction.

However, notwithstanding Anki, the privacy and data-protection history of smart toys are not good – exemplified by My Friend Cayla (Genesis Toys 2014–17), which recorded and relayed secrets told by the child to Cayla to both Genesis Toys and Nuance, the firm that handled Cayla's natural language processing capabilities. In 2017, a German regulator banned Cayla, describing it as a 'concealed surveillance device' (Bundesnetzagentur 2017). Hacks and exposure of weak security plague connected and smart toys (Haynes et al. 2017). As a result, in addition to scope for data protection, security and privacy harms, there are other potential grievances, such as betrayal, violation and breach of trust between child and technology. That this occurs with children raises unique concerns because they are typically more trusting and cannot be expected to have the same level of awareness as adults. At the level of ethics, this chapter avoids reactionary moralizing, preferring to inquire into the terms of the human-agent interaction, rather than whether children should be able to use novel modes of engagement through emotoys. As such, it resists both 'overly optimistic or dystopian predictions of the near future of children's playful techno-culture' (Giddings 2019: 70), focusing instead on the terms of connection and engagement. For the child, these are experiential and developmental but structured by questions of law, consumer protection, commercial logic, data security and privacy. Nevertheless, although this chapter prefers specifics over giddy optimism or bleak dystopia, emotoys represent something historically novel. Crudely, toys may be split into three stages: (1) traditional toys, which are items upon which a child's feelings, emotions and experienced/imagined situations are projected; (2) smart toys from the 1990s onwards, such as Furby and Tamagotchi, that involve awareness, monitoring and care of the toy's emotional state and (3) emotoys, from the 2010s onwards, that watch us back.[2] This requires careful consideration for how emotoys are designed and regulated, especially because the type of empathy and emotional sensitivity displayed is of a weak and ultimately uncaring sort.

POLICY AND RELATIONSHIPS

There are clear ethical, human rights and legal questions to be asked of emotoys. At an international level, the United Nations Convention on the Rights of the Child (OHCHR 1990) has a keen interest in the children's best interests, development, mental wellbeing, privacy, liberty, rest, leisure, play, recreational activities and the right to be protected from economic exploitation (see Articles 3, 5, 16, 18, 27, 29, 31 and 32). In Europe, there is broad coverage of child rights and wellbeing,[3] with the European Union's General Data Protection Regulation (GDPR) highlighting that special protections are necessary for children (Recital 38), especially regarding consent. Where children are below 16 years (but not lower than 13), consent must be given by people with parental responsibility over the child (Article 8). Data about children raises questions of how to render complex technical processes in plain terms so that older children and parents may understand. As ever, there is the question of whether consent is meaningful, informed and truly voluntary. Advising on consent, connected toys and devices, the UK's Information Commissioner's

Office states that customers should be able to view privacy information, terms and conditions of use and other relevant information online without having to purchase and set up the device first, so that they can make an informed decision about whether or not to buy the device in the first place' (ICO 2020). Other GDPR issues include data minimization, meaning that the personal data collected is 'adequate, relevant and limited to what is necessary in relation to the purposes for which they are processed' (Article 5(1c)). Whereas the GDPR provides an *overall* data-protection framework for the protection of personal data, the proposed e-Privacy Regulation pays closer attention to privacy protections, confidentiality of communications and fundamental rights and freedoms, which may also contain non-personal data and data related to a legal person. Interestingly, the e-Privacy Regulation does not explicitly mention children (Milkaite and Lievens 2019). Also, emotion is not mentioned in GPDR, but it is mentioned in Recital 2 and 20 of the proposed e-Privacy Regulation, but not the Articles themselves (recitals are the reasons for articles, which are the rules and stipulations).

As emotoys become more sophisticated than Cozmo, there are wider child wellbeing issues at stake than found in law, not least child emotional and psychological development (theory of mind), engagement with synthetic personalities, animism, relationships and attachments and related concerns. These are potentially captured in aforementioned articles in the UN Convention on the Rights of the Child (on child development, personality development, parental and carer responsibilities, rest and leisure and play and recreation activities), but the rights were not designed with these issues in mind. Novel forms of interactivity raise questions of meaning, significance and the nature of relationships. Fundamentally, it is the nature of connection between people and things. This brings to mind Sherry Turkle's observation from the 1980s that 'Some children think computers are alive, some think they are not, others settle on "sort of"' ([1984] 2005: 52). Echoing Weizenbaum's (1976) 'Eliza', the anthropomorphized computer program that played the role of an Rogerian psychiatrist, Westlund and Breazeal (2015), for example, report on human–robot interaction 'Wizard-of-Oz' studies of children telling secrets to robots (because they will not tell adults), deception about the robot's autonomy and agency, the importance of trust in a relationship and children's general understanding of robots. One of their key findings was that even though children were shown how their robots worked (thus breaking the deception), they continued to behave as if the robot was a separate entity.

Power asymmetry between corporations and citizens must also be avoided in relation to other data personalization harms. Bertolini

(2018), for example, signals key problems with deception in the design of robots that seek to influence perceptions of animacy. Distinguishing between nature and appearance, Bertolini cites Cameron et al. who point to:

> *morphology* (whereby a lifelike humanoid form induces the expectations of more sophisticated interaction and leads children to frame these applications into a distinct category of person-machine hybrid); *responsiveness* to the environment in a way that is perceived as being socially adequate and hence meaningful to the user; *movement* that is perceived as to be autonomous and goal oriented. (2018: 652, original emphasis)

The critical concern is exploitation due to reward mechanisms and undeveloped theories of mind that may be abused by vendors of robots and/or synthetic personalities, especially as children are deceived into design illusions of animism and begin to share secrets, confidence, other intimate interaction and potentially become significantly emotional attached. A key aspect of the question of deception is *patiency* in emotoys[4] or the extent to which children as living subjects are an 'other' for the toys that is doing the empathizing. While emotoys may simulate the capacity for conscious and moral orientation to the child user, there is at worst deception and at best wilful participation by the child in their own seduction (the distinction thus being one agency and awareness). The concern about underdeveloped theory of mind and scope for deception is amplified by the business failure of Cozmo that arguably failed because of the lack of a data-oriented/platform revenue model that would have sought to commodify data about children.

Also of ethical note is whether deception and manipulation *are* acceptable, if the child's interests are being served, such as to discover cases of abuse that a child feels that they cannot confide to a good and responsible adult (such as a parent or teacher). Bertolini's view is that a consequentialist perspective (where ends justify the means) is not valid because deception with intent is innately unethical, but also that there are as yet unknown consequences of deceiving and manipulating children. However, early findings about the ontological status of synthetic personalities and emotoys are not consistent. Lovato et al. (2019), for example, find that children aged 5–6 years are sceptical of voice assistants and creative in their attempts to test how reliable the gadgets are. Answers such as 'I don't know' to 'Do unicorns exist?' made them appear less trustworthy to children. While children may or may not be duped by synthetic personalities in emotoys, caution is required that such findings are not used to justify outsourcing governance [self] responsibility to children and pressured parents.

As emotional AI, domestic empathic technologies and emotoys develop, it is likely that a challenge will be how to make policy regarding engagement with synthetic personalities when (1) their sophistication is a matter of degree and (2) and child technological literacy varies between 0 and 16. Good policy is rarely made in response to moral panic, so it is advisable that a wait-and-see approach is adopted to any extra policy in addition to existing law, but this is not to suggest regulatory complacency. Emotoys *will* develop in cleverness, and the scope for privacy, relational, developmental and other child harms is clearly foreseeable. Other harms are a little more oblique, but potentially important if they come to pass. A key one is the issue of 'reflexive intentionality' (Rosenberger and Verbeek 2015: 22) in relation to emotion, empathic technologies and synthetic personalities. To simply put, this refers to the experience of being surveilled and objectified by sensing objects (such as emotoys). Certainly this is a privacy concern, but it is also one about socialization and acculturalization to the logics of our technologies. Will children begin to perform for the programmed expectations of emotoys? Will they alter emotional expressions in play to conform to the programmed psychological models, such as basic emotions? Will they amend speech behaviour to improve understanding by emotoys with voice abilities, or will they self-censor their speech? Further, on being sensed, *if* deemed at all acceptable, at what ages are children able to comprehend empathic objectification, and what alterations to their behaviour (through sensing and interaction) are acceptable?

Last, with today's children being tomorrow's adults, what sort of culture do emotoys engender? If reflexive intentionality (the sense of being watched and objectified) is normalized in childhood, what does this mean for later years? Connecting back with this book's core interest, intimacy and technology, we suggest that domestic reflexive intentionality is likely to remain an issue. These technologies are not like other quiescent domestic IoT objects, such as toasters and refrigerators. Emotoys of child and adult form will monitor and report on highly intimate dimensions of human life. This brings significant privacy issues into the equation and, for child and adult emotoys alike, questions also of risk and ambivalence. Are these new forms of pleasure a Faustian bargain?

Acknowledgement

This work is supported by Engineering and Physical Sciences Research Council (grant number EP/R045178/1).

References

Aloufi, Ranya, Haddadi, Hamed and Boyle, David (2019), 'Emotionless: Privacy-preserving speech analysis for voice assistants', *arXiv*, https://arxiv.org/abs/1909.08500. Accessed 3 February 2020.

Altenberger, Felix and Lenz, Claus (2018), 'A non-technical survey on deep convolutional neural network architectures', *arXiv*, https://arxiv.org/abs/1803.02129. Accessed 3 February 2020.

Anki (2019), 'Cozmo', https://anki.com/en-gb/cozmo/product-details.html. Accessed 3 February 2020.

Barrett, Lisa F., Adolphs, Ralph, Marsella, Stacy, Martinez, Aleix M. and Pollak, Seth D. (2019), 'Emotional expressions reconsidered: Challenges to inferring emotion from human facial movements', *Psychological Science in the Public Interest*, 20:1, pp. 1–68.

Bertolini, Andrea (2018), 'Human-Robot interaction and deception', *Osservatorio del diritto civile e commerciale*, 2, pp. 645–59.

Breazeal, Cynthia (2002), *Designing Sociable Robots*, Cambridge, MA: MIT Press.

Bundesnetzagentur (2017), 'Bundesnetzagentur removes children's doll "Cayla" from the market', https://www.bundesnetzagentur.de/SharedDocs/Pressemitteilungen/EN/2017/170220 17_cayla.html. Accessed 3 February 2020.

Druga, Stefania, Breazeal, Cynthia, Williams, Randi and Resnick, Mitchek (2017), '"Hey Google is it OK if I eat you?" Initial explorations in child-agent interaction', https://drive.google.com/file/d/0B552cSaArazGUGpma2RKWD-FTYUU/view. Accessed 03 February 2020.

Ekman, Paul and Friesen, Wallace V. (1971), 'Constants across cultures in the face and emotion', *Journal of Personality and Social Psychology*, 17:2, pp. 124–29.

Ekman, Paul and Friesen, Wallace V. (1978), *Facial Action Coding System: A Technique for the Measurement of Facial Movement*, Palo Alto: Consulting Psychologists Press.

ePrivacy Regulation (ePR) (2017), *Proposal for a Regulation of the European Parliament and of the Council Concerning the Respect for Private Life and the Protection of Personal Data in Electronic Communications and Repealing Directive*, (2017) 2002/58/EC (Regulation on Privacy and Electronic Communications) Brussels, 10.1.2017 COM 10 final 2017/0003 (COD), https://eur-lex.europa.eu/legal-content/EN/TXT/?uri=CELEX%3A52017PC0010. Accessed 3 February 2020.

General Data Protection Regulation (GDPR) (2016), Regulation (EU) 2016/679 of the European Parliament and of the Council of 27 April 2016 on the Protection of Natural Persons with Regard to the Processing of Personal Data and on the Free Movement of Such Data, and Repealing Directive 95/46/EC. https://publications.europa.eu/en/publication-detail/-/publication/3e485e15-11bd-11e6-ba9a-01aa75ed71a1/language-en. Accessed 3 February 2020.

Giddings, Seth (2019), 'Toying with the singularity: AI, automata and imagination in play with robots and virtual pets', in G. Mascheroni and D. Holloway (eds), *The Internet of Toys: Practices, Affordances and the Political Economy of Children's Smart Play*, London and New York: Palgrave MacMillan, pp. 67–87.

Goldman, Alvin I. (2008), *Simulating Minds: The Philosophy, Psychology, and Neuroscience of Mindreading*, Oxford: Oxford University Press.

Gunkel, David J. (2017), *The Machine Question: Critical Perspectives on AI, Robots, and Ethics*, Cambridge, MA: MIT Press.

Haynes, Jeffrey, Ramirez, Maribette, Hayajneh, Thaier and Bhuiyan, Md Zakirul A. (2017), 'A framework for preventing the exploitation of IoT smart toys for reconnaissance and exfiltration', in G. Wang, M. Atiquzzaman, Z. Yan and KK Choo (eds), *Security, Privacy and Anonymity in Computation, Communication and Storage*. SpaCCS 2017, vol. 10658, Cham: Springer, pp. 581–92.

Husserl, Edmund (1980), *Phenomenology and the Foundations of the Sciences*, The Hague: Martinus Nijhoff.

ICO (2020), *Connected Toys and Devices*, https://ico.org.uk/for-organisations/guide-to-data-protection/key-data-protection-themes/age-appropriate-design-a-code-of-practice-for-online-services/14-connected-toys-and-devices. Accessed 3 February 2020.

Jones, Meg L. and Meurer, Kevin (2016), 'Can (and should) Hello Barbie keep a secret?', *2016 IEEE International Symposium on Ethics in Engineering, Science and Technology*, Vancouver, BC, Canada, 13–14 May 2016, https://papers.ssrn.com/sol3/papers.cfm?abstract_id=2768507. Accessed 3 February 2020.

Kory-Westlund, Jaqueline M. and Breazeal, Cynthia (2015), 'Deception, secrets, children, and robots: what's acceptable?', https://www.media.mit.edu/publications/deception-secrets-children-and-robots-what-s-acceptable. Accessed 3 February 2020.

Lipton, Zachary C., Berkowitz, John and Elkan, Charles (2015), 'A critical review of recurrent neural networks for sequence learning', *arXiv*, https://arxiv.org/abs/1506.00019. Accessed 3 February 2020.

Lovato, Silvia B., Piper, Anne M. and Wartella, Ellen A. (2019), ''Hey Google, do unicorns exist?': Conversational agents as a path to answers to children's questions', *ACM*, https://www.scholars.northwestern.edu/en/publications/hey-google-do-unicorns-exist-conversational-agents-as-a-path-to-a. Accessed 3 February 2020.

Mascheroni, Giovanna and Holloway, Donell (2019), 'Introducing the internet of toys, in G. Mascheroni, and D. Holloway (eds), *The Internet of Toys: Practices, Affordances and the Political Economy of Children's Smart Play*, London and New York: Palgrave MacMillan, pp. 1–22.

McDuff, Daniel (2014), *Crowdsourcing Affective Responses for Predicting Media Effectiveness*, Ph.D. Thesis, http://affect.media.mit.edu/pdfs/14.McDuff-Thesis.pdf. Accessed 3 February 2020.

McDuff, Daniel and Czerwinski, Mary (2018), 'Designing emotionally sentient agents', *Communications of the ACM*, 61:12, pp. 74–83, http://dx.doi.org/10.1145/3186591. Accessed 3 February 2020.

McDuff, Daniel, Gontarek, Sarah and Picard, Rosalind (2014), 'Remote measurement of cognitive stress via heart rate variability', https://www.researchgate.net/publication/270658751_Remote_measurement_of_cog/nitive_stress_via_heart_rate_variability. Accessed 3 February 2020.

McStay, Andrew (2016), 'Empathic media and advertising: Industry, policy, legal and citizen perspectives (the case for intimacy)', *Big Data & Society*, 3:2, pp. 1–11.

McStay, Andrew (2018), *Emotional AI: The Rise of Empathic Media*, London: Sage.

McStay, Andrew and Urquhart, Lachlan (2019a), "This time with feeling?" Assessing EU data governance implications of out of home appraisal based. Emotional AI', *First Monday*, https://firstmonday.org/ojs/index.php/fm/article/view/9457/8146. Accessed 3 February 2020.

McStay, Andrew and Urquhart, Lachlan (2019b), 'Emotional AI and EdTech: Serving the public good?', *Learning Media & Technology*, https://www.tandfonline.com/doi/full/10.1080/17439884.2020.1686016. Accessed 3 February 2020.

McStay, Andrew and Urquhart, Lachlan (2020), 'Emotional AI, soft biometrics and the surveillance of emotional life: an unusual consensus on privacy', *Big Data & Society*, pre-publication.

Milkaite, Ingrida and Lievens, Eva (2019), 'The internet of toys: Playing games with children's data?', in G. Mascheroni and D. Holloway (eds), *The Internet of Toys: Practices, Affordances and the Political Economy of Children's Smart Play*, London and New York, Palgrave MacMillan. pp. 285–305.

Office of the United Nations High Commissioner for Human Rights (1990), *Convention on the Rights of the Child*, https://www.ohchr.org/Documents/ProfessionalInterest/crc.pdf. Accessed 3 February 2020.

Picard, Rosalind W. (1997), *Affective Computing*, Cambridge, MA: MIT.

Picard, Rosalind W. (2007), *Toward Machines with Emotional Intelligence*, http://affect.media.mit.edu/pdfs/07.picard-EI-chapter.pdf. Accessed 3 February 2020.

Ren, Shaoqing, He, Kiaming, Girshick, Ross and Sun, Jian (2015), 'Faster R-CNN: towards real-time object detection with region proposal networks', *arXiv*, https://arxiv.org/abs/1506.01497. Accessed 3 February 2020.

Rosenberger, Robert and Verbeek, Peter-Paul (2015), 'Introduction', in R. Rosenberger and P-P. Verbeek (eds), *A Field Guide to Postphenemenology*, London: Lexington Books, pp. 9–41.

Russell, James A. (1994), 'Is there universal recognition of emotion from facial expression? A review of the cross-cultural studies', *Psychological Bulletin*, 115:1, pp. 102–41.

Schleifer, Theodore (2019), 'The once-hot robotics startup Anki is shutting down after raising more than $200 million', *Recode*, https://www.vox.com/2019/4/29/18522966/anki-robot-cozmo-staff-layoffs-robotics-toys-boris-sofman. Accessed 3 February 2020.

Turkle, Sherry (1984 [2005]), *The Second Self: Computers and the Human Spirit*, Cambridge, MA: MIT.

Turkle, Sherry (2011), *Alone Together: Why We Expect More From Technology and Less From Each Other*, New York: Basic Books.

Viahos, James (2019), *Talk to Me: How Voice Computing Will Transform the Way We Live, Work, and Think*, New York: Houghton Mifflin Harcourt.

Statista (2019), 'Size of the sex toy market worldwide from 2019 to 2026', https://www.statista.com/statistics/587109/size-of-the-global-sex-toy-market/. Accessed 3 February 2020.

Weizenbaum, Joseph (1976), *Computer Power and Human Reason: From Judgment to Calculation*, San Francisco: W. H. Freeman.

Winfield, Alan (2012), *Robotics: A Very Short Introduction*, Oxford: Oxford University Press.

Zelenski, John M. and Larsen, Randy J. (2000), 'The distribution of basic emotions in everyday life: A state and trait perspective from experience sampling data', *Journal of Research in Personality*, 34:2, pp. 178–97.

FIGURE 4.2 (pages 82–83): !Mediengruppe Bitnik, *Ashley Madison Angels at Work in London*, 2017. Five channel video installation, full-HD, 16:9, sound, 40" LCD screens, trolley stands, video players, cables, pink neon light. Courtesy of the artists and Annka Kultys Gallery, London.

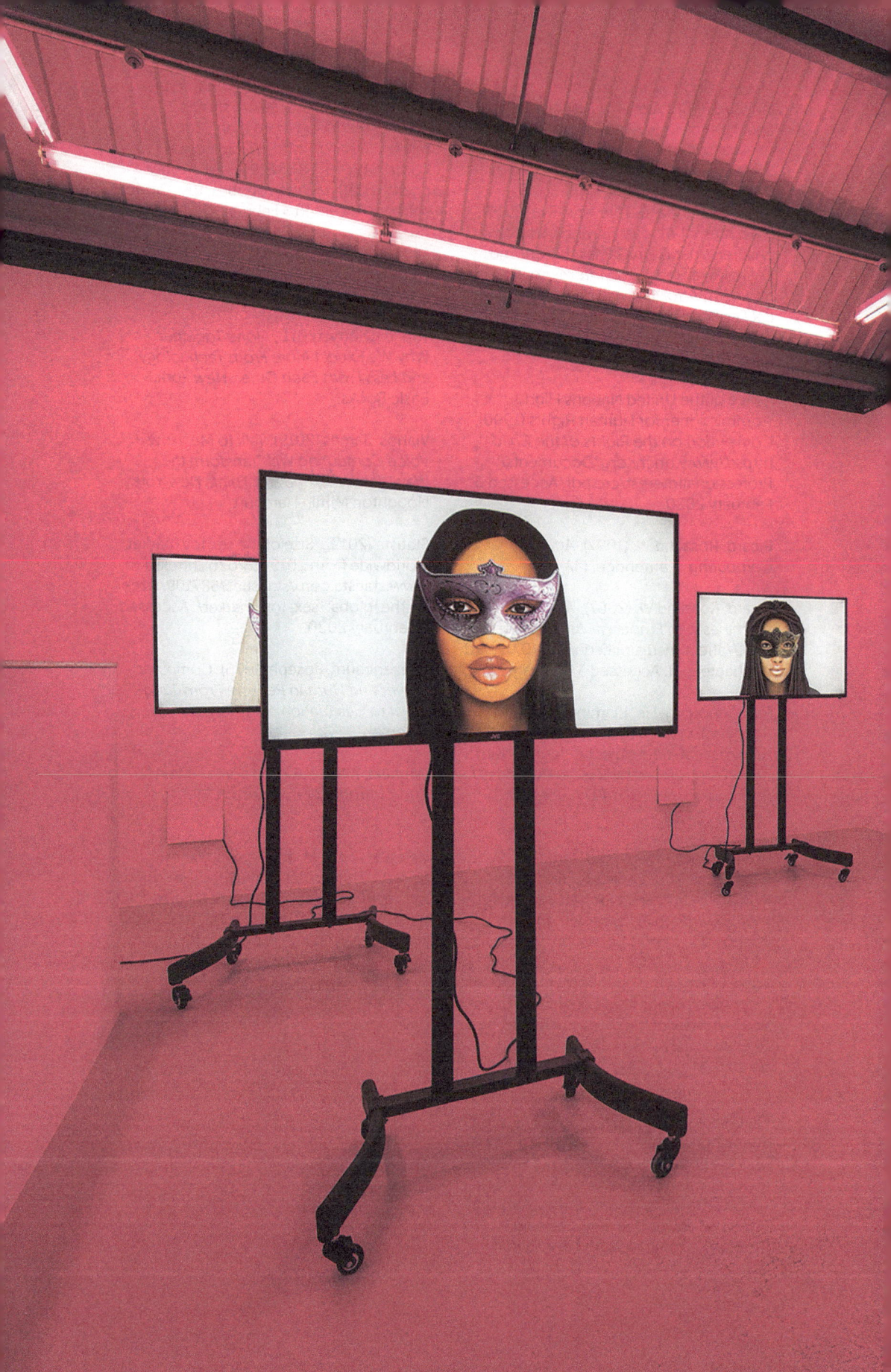

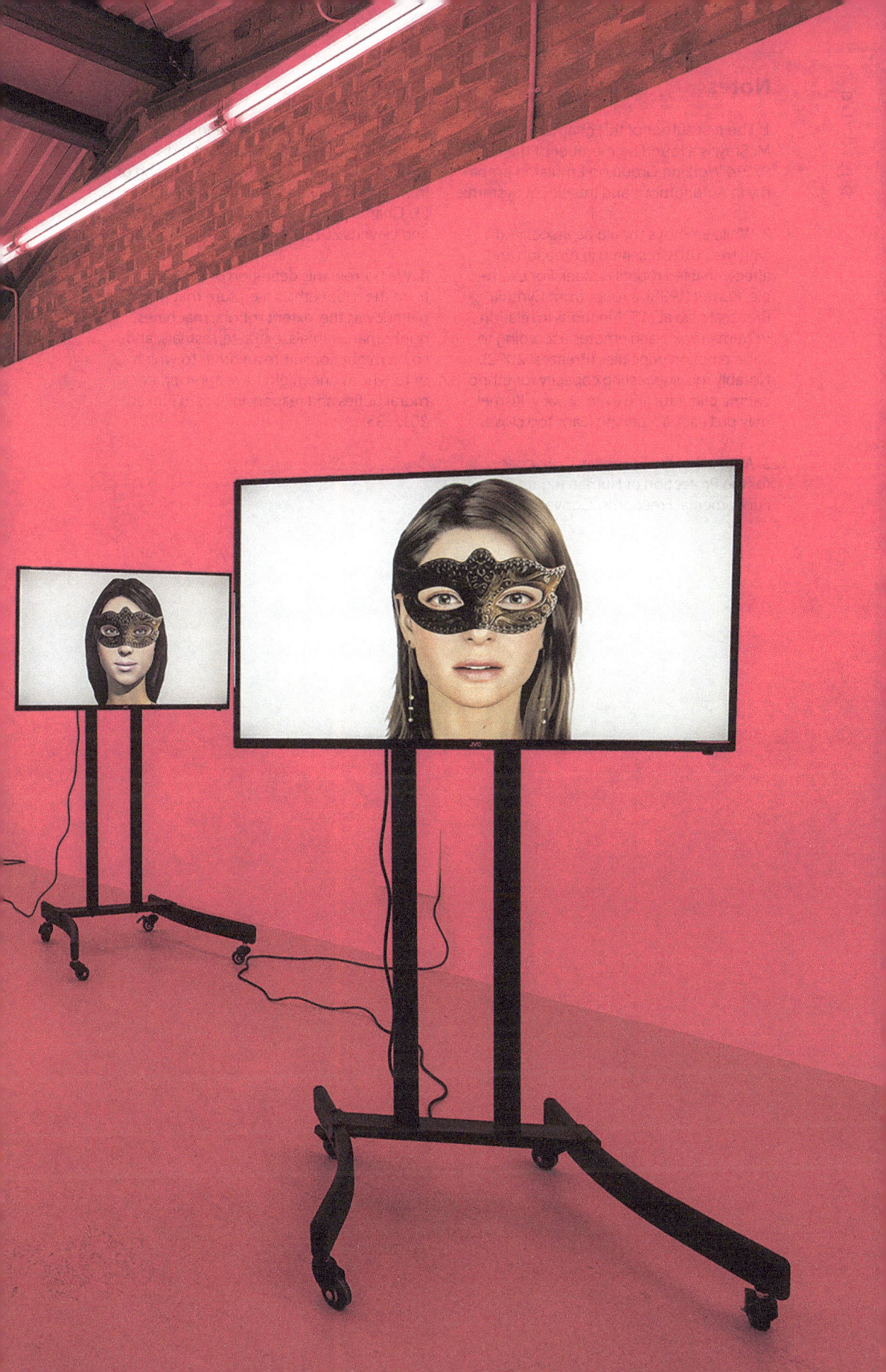

Notes

1. The first author of this chapter, Andrew McStay, is a founding member of the IEEE P7014 Working Group on Emulated Empathy in Autonomous and Intelligent Systems.

2. While emotoys should be associated with the 2010s, research in emotion and affect-enabled robots is older. For example, *Kismet* (1998), a robot from Cynthia Breazeal's lab at MIT, functions in relation to human voice and emotes according to basic emotion principles (Breazeal 2002). Notably too, suggesting capacity for ethnocentric etiquette and even privacy, Kismet may pull back if a person leans too close.

3. Article 8 of the European Convention for the Protection of Human Rights and Fundamental Freedoms; Convention 108+ of the Council of Europe Modernised Convention for the Protection of Individuals with Regard to the Automatic Processing of Individual Data; and articles 7 and 8 of the EU Charter of Fundamental Rights (Milkaite and Lievens 2019).

4. We borrow this definition of patiency from the robot ethics literature that moral patiency as the 'extent robots, machines, nonhuman animals, extra-terrestrials, and so on might constitute an *other* to which or to whom one might have appropriate moral duties and responsibilities' (Gunkel 2017: 93).

5
Self-Fashioning Desire

Social media is a big scene for self-advertising. You can enter this scene from your phone (or any cellular or WiFi-tethered deceive) and decorate it with props that are within a click's reach. Used for professional and private purposes by millions of people of different age and strata, social media are still the favourite and most natural playground for digital natives who use it to find friendship, love, work and identity. As a rather unconstrained and yet distancing medium, social platforms (Facebook, Twitter, Instagram, etc.) invite behaviours that we normally do not exhibit in person-to-person live interactions. They also invite radical forms of conduct organized by two tendencies: exhibitionism and bullying. The former relates to what researchers term the *narcissification of the self* – a compulsive need to share one's filtered and well-stylized (often themed) persona(s). The latter refers to uninhibited violence, visible in the language of tweets, texts and nudges that transform daily shares into attacks violating privacy and dignity – attacks that for many young adults is a way of fashioning their social presence.

Kill Your Darlings (2012) is an artwork by Jeroen van Loon that depicts both the tendencies in an installation of small LCD-keychains shaped in the form of little hearts that display Twitter messages of teenage girls combined with their profile pictures. The images ooze young feminine cuteness tailored to the demand of social profiling. The messages, on the other hand, articulate insult, hatred, contempt and abuse, enunciating the new standard of 'social bitchiness', the key idiom of which is, 'I hope you die alone, with 72 cats, bitch!'

Kill Your Darlings documents the transformation of young women as 'beautiful little darlings' into 'cruel little devils'. It also documents the transformation of a social space under digitalism, showing how online cultures have changed our understanding of privacy, media and social limits. The virtual space where we broadcast our individual opinions and lifestyles to the world is also a space of large audiences. It is also a space of anonymous and less anonymous meet-ups, which test the boundaries of our social capacity and wherein 'I am what I share' is the primary lingo of self-presentation.

Are we what we share online? The investment into self-profiling we witness in our daily scroll-downs proves that a post might be a new 'cogito' machine. Fuelled by the whirlpools of looks, likes and comments, the machine spins into us the complete ignorance of the other, to let the other in us launch itself to the world.

Kill Your Darlings

JEROEN VAN LOON

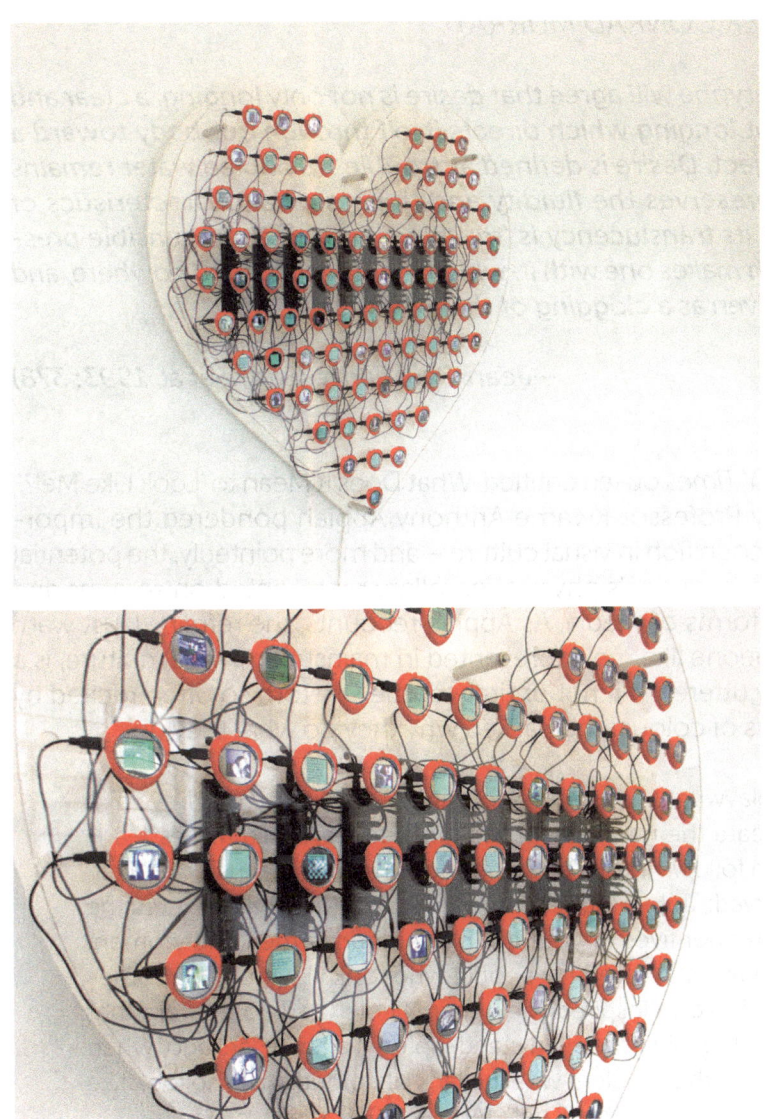

FIGURE 5.1–5.2: Jeroen van Loon,
Kill Your Darlings, 2012. Installation,
97 LCD displays, 10 USB hubs, wood,
plexiglass, 120 cm x 120 cm x 18.8 cm.
Courtesy of the artist.

The Greatest Love of All: Recognition, Self-Love and the Imaging of Desire

DEREK CONRAD MURRAY

In fact, everyone will agree that desire is not only longing, a clear and translucent longing which directs itself through our body toward a certain object. Desire is defined as trouble... troubled water remains water; it preserves the fluidity and the essential characteristics of water; but its translucency is 'troubled' by an inapprehensible presence which makes one with it, which is everywhere and nowhere, and which is given as a clogging of the water itself.

—Jean-Paul Sartre (Sartre et al. 1993: 378)

In a 2019 *NY Times* op-ed entitled 'What Does It Mean to 'Look Like Me'?', philosophy Professor Kwame Anthony Appiah pondered the importance of recognition in visual culture – and more pointedly, the potential impact of seeing one's physical corollary represented on screens and in popular forms of media. As Appiah recounts, the refrain, 'I just want to see someone like me' represented in mainstream visual culture, is a commonly uttered – if not entirely cliché – retort, so often recited by entertainers of colour when asked why they do what they do:

> The playwright Tarell Alvin McCraney, explaining what drove him to create the new television drama series 'David Makes Man', which follows the life of a black boy in a public-housing project, observed: 'John Hughes made several movies that depicted the rich interior lives of young White American men and women. I just want the same for people who look like me'. The comedian Ali Wong inspired the writer Nicole Clark to confess that she 'didn't think she liked stand-up until a few years ago, when I realized the problem was the lack of comedians who look like me and tell jokes that I 'get'. (Appiah 2019: n.pag.)

Appiah is right to point out that this ever-popular notion alludes to what the philosopher calls a 'kinship of social identity': It functions as a means to express in-group cultural commonalities (Appiah 2019). Though it may speak to a desire for recognition that is easily articulated, we must also acknowledge that group identity (and the solidarities it calls for) has always been more aspirational than a tangible reality. Appiah's consideration of this representational problem argues that we may need to complicate what it means to see people who 'look like me', while acknowledging the fallaciousness and impossibility for images to capture the complexities and ideological slippages of identity.

The aspirational dimensions of recognition in the visual realm, of which Appiah speaks, resonates in many ways with self-imaging online, most commonly articulated in the visual form known as the selfie or digital self-portrait. The now ubiquitous term has become synonymous with our twenty-first century techno-crazed society, even while the act itself has been maligned as a puerile and largely vacuous visual expression. Common on social media sites like Facebook and Instagram, the selfie has become a powerful means of self-expression, encouraging its makers to share the most intimate and private moments of their lives. The popularity of self-imaging online is unprecedented, yet within popular journalism the selfie is reviled and regarded as a shallow expression of online narcissism. I have written previously about the more political and socially engaged dimensions of online self-imaging while considering the gendered implications of both the act and its common (mis)characterization as a largely female-driven form of communication (Murray 2015: 490). While many of the more radical mobilizations of the selfie are produced by women, it is not, contrary to dominant discussions, a uniquely female act. Despite that fact, popular discourses have largely characterized the selfie as a symptom of a narcissistic society – or more accurately, is the assertion that those who make selfies are themselves suffering from mental health disorders (Murray 2020: 21).

In response to the aforementioned tendency, this chapter argues for a reassessment of digital culture's impact/s on contemporary life while considering how certain constituencies have responded to the increasing presence of new technologies. I acknowledge that the utopian promise of digital culture has, in many respects, contributed to the dismantling of democracy and is complicit in the rise of inequality and the erosion of privacy. By exploring the impact of digital utopianism in the visual realm, this research explores how young women of colour, in particular, have responded to the ubiquity of new technologies as means to assert a sense of personal value, recognition and empowerment. The representational expressions of marginalized groups online have often functioned as a counterpoint to the dominant culture's tendency towards devaluing, misrepresentation and erasure – though these expressions have nonetheless absorbed digital mediums, making productive use of their participatory networks and interactive platforms. One of the more generative aspects of self-imaging online is its engagement with new forms of image making in the twenty-first century, that utilize (and often weaponize) individuated, participatory and interactive practices as a powerful means for social engagement. Ultimately, the aim is to explore the impact of digital technology on contemporary life from an anti-utopian perspective that is lucid about the implicit identity politics expressed within the often-fraught and misinterpreted act of online self-imaging.

The critical framework for this book endeavours to explore love and intimacy under digitalism while unpacking such themes as practicing romance, togetherness – as well as notions of fantasy, fetish and eroticism. In contemplating this thematic under the banner of 'Self Fashioning Desire', I tend to think of love and intimacy (and the self-image) in terms of desiring the self. Moreover, I critically position the selfie as one of intense intimacy: as a form of self-fashioning that is often about attempting to assuage feelings of lack, isolation and alienation – as a means to *love oneself*. For that reason, this chapter takes a more speculative approach that investigates the self-imaging of women of colour, not as an effort to generate external sexual desire (something that beckons), but rather to create a framework in which they begin to desire themselves: to create a representational system of value for subjectivities who experience persistent devaluing and/or erasure – a phenomenon that we might regard as a *romance with the self*. As mentioned, these expressions flourish under a constant pejorative threat of ridicule and pathologizing rhetoric. For that reason, there is a great need to understand the logics that drive the ubiquity of self-imaging in online cultures as a tool for self-definition.

In 2013, *The Oxford English Dictionary* announced 'selfie' as the word of the year – a distinction that has inaugurated its introduction into the public consciousness. A selfie has been defined as a photograph that one has taken of oneself, typically one taken with a smartphone or webcam and uploaded to a social media website. Despite the increasing ubiquity of the term – and the popularity of the act itself – the tendency has been to position online forms of self-imaging as culturally corrosive, pathological and even mortally dangerous. As a result, the selfie debate has evolved from a lighthearted discussion about the perils of technology and consumption to the pathologizing of the image maker. The aim here is to give insight into a very contemporary discussion about the impact of technology and social media as a means to disseminate and share images. The term 'selfie', in its popular usage, delineates a very particular engagement with technologies of image making, a phenomenon that has led to public debate about the potential corrosive effects of technology on our individual and collective selves.

There is, on the other hand, a continued need to explore the political, ideological and aesthetic complexity at the heart of the selfie phenomenon and contemplate whether the urge to compulsively self-image in the twenty-first century is mere narcissism or if it holds the potential for more redemptive meanings. Moreover, there is a significant amount of serious scholarship being done on the subject of digital self-portraiture (primarily in the social sciences and communications), but most of it is *not* engaged with the visual. While there is a need for scholarship that is innovative, knowledgeable and insightful about online self-imaging as a social, economic and technological phenomenon, there is a paucity of research that is engaged with digital self-portraiture *as* representation. Scholars Edgar Gómez and Ellen Thornham make a contrasting argument, suggesting that placing undue attention on self-representation is to ultimately miss the point:

> We argue that contemporary understandings of selfies either in relation to a "documenting of the self" or as a neoliberal (narcissistic) identity affirmation are inherently problematic. Instead, we argue that selfies should be understood as a wider social, cultural, and media phenomenon that understands the selfie as far more than a representational image. This, in turn, necessarily redirects us away from the object "itself," and in so doing seeks to understand selfies as a socio-technical phenomenon that momentarily and tentatively holds together a number of different elements of mediated digital communication. (Gómez et al. 2015: 1)

Gómez and Thornham's position is a widely held one across the spectrum of social science research – even while the selfie, as a highly personalized representational act, speaks directly to discourses around recognition, as well as to the *kinship of social identity* of which Appiah speaks: that pesky need to *see oneself* imaged.

Needless to say, the interpretive quandary posed by these contrasting approaches alludes quite troubling towards the relation between the individual consumer and the rapacious forces of techno-capitalism that bear down upon them. Scholar Henry Giroux compellingly argues that 'freedom has become an exercise in self-development rather than social responsibility'; further intimating that the tendency towards neocapitalist privatization and voracious consumerization has served to erode the public good:

> American society is in the grip of a paralyzing infantilism, marked by a crisis of history, memory, and agency. Everywhere we look, the refusal to think, interrogate troubling knowledge, and welcome robust dialogue and engaged forms of pedagogy are now met by the fog of rigidity, anti-intellectualism and a collapse of the public into the private. A politics of intense privatization and its embrace of the self as the only viable unit of agency appears to have a strong grip on American society, as can be seen in the endless attacks on reason, truth, critical thinking and informed exchange, or any other relationship that embraces the social and the democratic values that support it. (Giroux 2015: 156)

Giroux tends to view the selfie phenomenon as a troubling sign that 'a vision of the good society has now been replaced with visions of individual happiness characterized by an endless search for instant gratification and self-recognition' (Giroux 2015: 156). He further states that 'the personal appears to be the only politics that matters in providing both emotional gratification and a tangible referent for negotiating social problems' (Giroux 2015: 156). Giroux's central concern here is the decimation of individual privacies, engendered by the state and the corporate sphere. He argues quite compelling that there is an 'increasing view of privacy on the part of the American public as something to escape from rather than preserve as a precious political right' (Giroux 2015: 156). To that point, the willingness of consumers to give-over their most personal information is a hallmark of the social media phenomenon: a feature of twenty-first century life that, as Giroux laments, acculturates the masses into the intrusion of consumer-based surveillance practices. There is really no productive counterpoint to this argument, as we come to grips with the manner in which technology has created new and more invasive means to exert its control. The scholar acknowledges that while the selfie does not offer-up the types of information that is most concerning, it does, in fact, transform the self into a matter of public concern (Giroux 2015: 156). 'Privacy has mostly become synonymous with a form of self-generated, non-stop performance – a type of public relations in which privacy is valued only for the way it makes possible the unearthing of secrets, a cult of commodified confessionals and an infusion of narcissistic, self-referencing narratives' (Giroux 2015: 157).

Many commentators have lamented what is perceived as a societal descent into self-obsession and the shallowness of individual posturing. And the selfie is the perfect foil for these concerns. We can, of course, utilize online self-portraiture and image sharing as a barometer for the hegemonies of state and corporate surveillance. Though perhaps, we need to look more closely, particularly as we acknowledge that throughout history, the tools meant to dominate have often been weaponized against those very forces. While not attempting to uncritically redeem the selfie phenomenon of its troubling dimensions, I argue that the personal is *still* political, even while the capitalist subject wrestles with the threats posed by techno-capitalist aggression.

THE SOCIAL MEDIA INFLUENCER AND THE RISE OF *INSTAFAME*

While I agree with the more pessimistic framings recounted above, there is something about the selfie that continues to tug at the cultural fabric: a quality that is perhaps more unruly, more intimate and profoundly psychological. Though I am also sceptical at the apparent critical discomfort with the personal nature of the selfie – and with the unavoidable psychic intimacies, desires and longings expressed within the gesture to self-document. It is, in fact, a deeply personal act – one so often enacted in private and in moments of contemplative self-reflection. There hasn't been enough critical engagement with these expressed intimacies and longings, as well as the aesthetic peculiarities of the images. Selfies are disturbing images because they make a private moment public, which perhaps imbues them with a pornographic quality: a psychic dimension that is more exhibitionist and indulgent in spectacle, than violently voyeuristic.

In other words, something is given, more than taken or extracted. The differences here matter, and if for no other reason than we must engage with the notion of desire itself: desire for recognition, desire for visibility and self-worth or perhaps a desire to obtain simple humanity.

In Appiah's treatise on recognition, he cites the evolving public discourse in response to the popular Hollywood film *Crazy Rich Asians* – speaking directly to its impact on Asian viewers who rarely see actors who 'look like them' on the silver screen:

> *Crazy Rich Asians*, for all its shrewd social observations, is about group of people who are anything but representative. It's no criticism to say that a story in which the American daughter of a single working-class mother is whisked away by a billionaire to an enchanted kingdom of unfathomable richesse in Singapore has the same realism level as *The Princess Diaries*. What matters is that it's a Hollywood film about Asians in which Asians rule.

This has special significance, the writer Jiayang Fan says, when it comes to 'Asian-Americans, a largely made-up group that is united, more than anything else, by a historical marginalization.' (Appiah 2019: n.pag.)

While the film filled a representational void, it did not – in any realistic manner – represent Asian culture. On the other hand, it did indulge an *idea* of aspirational Asian-ness, regardless of its apparent perversities and representational limitations. It has been described in the press and on social as both a groundbreaking intervention in filmic diversity and nothing short of an Asian minstrel show. But neither of these characterizations interest me – rather, I am intrigued by the need for recognition in response to the film: that desire for reciprocity in the visual realm, so often denied. The impact of *Crazy Rich Asians*, specifically the notion that it represented a ground shift in Asian visibility and representation, is directly related to (if not a result of) the increasing presence of Asians on social media.

The rise of the social media influencer and online self-branding has emerged as a prominent feature of digital culture in the twenty-first century. Scholars Susie Khamis, Lawrence Ang and Raymond Welling define the social media influencer as engaged in the act of self-branding or personal branding, which 'involves individuals developing a distinctive public image for commercial gain and/or cultural capital' (Khamis et al. 2017: 191). Khamis et al. argue that central to this practice of self-branding is an individual who economically capitalizes on 'having a unique selling point, or a public identity that is singularly charismatic and responsive to the needs and interests of target audiences' (Khamis et al. 2017: 191). There is a concern that the human brand, which is essentially a personality with cultural capital, is the quintessential subject of neoliberal individualism at its most insidious. However, self-branding via social media is entirely based upon a desire for material gain, fame, celebrity and, most of all, recognition.

Khamis et al. point out that 'Social media is driven by a specific kind of identity construction – self-mediation – and what users post, share and like, effectively creates a highly curated and often abridged snapshot of how they want to be seen' (Khamis et al. 2017: 196).

This need for a kind of micro-celebrity has permeated all facets of online cultures, yet I am interested here in the evolution of selfie-taking (as an individualized and private act), into what we now know of as the influencer economy. Many young women, in particular, gained significant online following by taking selfies in their bedrooms and bathrooms – engaging in online conversations with their peers who were often great geographic distances from them. The selfie emerged as a kind of visual language that provided a forum to communicate and discuss a range of issues, from race, gender and sexuality, to body positivity, fashion and consumer habits. It was a logical progression that companies, both large and small, would take notice, eventually seizing on the popularity of private citizens who had unwittingly generated a significant viewer base. The enlisting of individuals with large social media following as brand ambassadors was therefore a logical progression.

The internet and especially the rise of social media has served to bolster the globalized reach of advanced computer capitalism – and the rise of self-branding has been particularly rapacious, encouraging an often-distorted notions of subjectivity and individual success that are subsumed by a consumerist logic. The gendered dimension of this phenomenon is particularly disturbing, particularly regarding young women, who often aggressively construct an online presence, with the intention of monetizing their personal image through self-branding. One needs to only gaze upon their surroundings, and they will surely see pairs of young women engaging in planned photoshoots, documenting their outfits and meals, in an effort to produce the ultimate vision of a chic and opulent existence. The social media app Instagram, with its massive user base, has created the possibility of achieving what is increasingly being labelled *Instafame*: a condition of social currency and economic opportunity, derived from garnering a large base of followers on the app. While there is a growing sense that influencers are exploited workers, toiling on a corporate leash – they often self-identify as an emergent creative class of content creators and cultural entrepreneurs. The truth might lie somewhere in the middle of these characterizations: that the influencer promotes brands in hopes of one day becoming one. The characteristics of this phenomenon are well known, specifically among youth cultures online that 'appear convinced that good looks, good living and conspicuous consumption (through artfully composed images of outfits, make-up, meals, holiday resorts, etc.) warrant adoration and emulation' (Khamis et al. 2017: 199).

This shift in celebrity culture has led to what Joshua Gamson argues is a 'heightened consciousness of everyday life as a public performance – an increased expectation that we are being watched, a growing willingness to offer up private parts of the self to watchers known and unknown, and a hovering sense that perhaps the unwatched life is invalid or insufficient' (Gamson 2011: 1061). Gamson's concern is the collision of a rapacious corporatized fixation with democratized fame, combined with an increasing adherence to the logics of self-surveillance. The fact that celebrity has been opened-up to a larger public, the influencer, Gamson argues, is pursued as entertainment as publicity technologies (Gamson 2011: 1069). The increasing digitization in the twenty-first century has created new and exciting mediums to generate meaning, which in turn has been particularly generative for minoritized groups.

The phenomena of online forms of entrepreneurialism, what Konrad Ng terms 'digital life', have 'become a compelling thread of the Asian American experience':

> Asian American culture has become a vibrant online popular culture with social consequence and creative possibility. Asian America is the most wired and engaged online community in the United States, and Asian Americans form a core constituency of America's "creative class" of professionals in the arts, sciences, and fields of technology, education, entertainment, and design. In sum, digital Asian America has become a cultural laboratory for popular meaning production and consumption; digital platforms have become the place to fashion a cultural economy with in-the-world, offline impact and activity. (Ng 2016: 139)

Despite his more optimistic view of Asian American digital participation, Ng reasons that we must resist the temptation to view this phenomenon through a neoliberal lens as 'evidence of some post-racial, multicultural internet in our democratic society. In other words, some might suggest that Asian Americans are the model minority of the web' (Ng 2016: 139). In contrast to this mischaracterization, Ng references the writings of Wendy Hui Kyong Chun, who argues for a more nuanced engagement with digital formations of racial identity, paying specific attention to how identity takes shape online (Chun 2012: 38). However, digital cultures have created the space for both self-representation and shared racial experiences for people of colour: constituencies who have been historically underrepresented by the various culture industries. Along those lines, Ng notes that digital representation for minority communities 'means wider opportunities for creative and critical expressions, new ways to manifest community and identity, and the chance to imaginatively engage with race in alternative spaces of meaning' (Ng 2016: 139). In sum, he argues that the 'Asian American thread in digital popular culture inflects digitalization discourses and the politics of race' (Ng 2016: 139). Ng offers some unique insights that help to illuminate how the politics of race may inform our understanding of the influencer phenomena. Asian influencers, in particular, have been both extremely active and influential in the evolution of creative expression online – often utilizing social media platforms to make powerful statements about, race and nationhood. Among these influencers is an apparent understanding that race, cultural identity and visual representation are inextricably bound.

The spectrum of Asian female influencers is incredibly diverse and encompasses a vast range of interests and industries – from life-style, wellness, travel and food cultures, to beauty and fashion. Among the notables are Chriselle Lim, Aimee Song, Nicole Warne, Margaret Zhang and Ellen V. Lora. It is worth noting that some of these influencers have emerged as major media figures, while others have parlayed their image into lifestyle and fashion brands. Chinese–Canadian Fashion and lifestyle blogger Vanessa Hong, for example, has garnered a significant following via her website *The Haute Pursuit*, as well as her Instagram account. Like many of her contemporaries, Hong's blogging straddles the worlds of self-help, wellness, fashion, beauty and travel – while self-consciously and meticulously constructing an image of professional and material success, health and style. In many respects, Hong's public persona is that of a fashion insider and life-style guru. This highly curated image expresses a range of desires and longing that have racial implications, not least the expressed fixation with European high culture as the barometer for social worth, access and acceptance. However, within these longings is also an acknowledgement of the persistent structural inequalities within the culture industries. The worlds of fashion, in particular, have historically resisted racial and ethnic diversity, so the apparent fascination (among female influencers of colour) with gaining access and mastering the industry's eccentricities is a common feature of their blogging output. Hong's aesthetic channels, described as fashion-forward cool, have been defined by its mini-malism. With her svelte frame and dyed blonde hair – a look that has become *de rigueur* among many Asian style and fashion bloggers – Hong's Instagram and website are filled with carefully constructed photographs of her in the act of strutting fashionably through the streets of Manhattan. These photographs are meant to replicate the types of images printed in the pages of fashion magazines like *Vogue*, featuring fashion designers, stylists, celebrities and various insiders attending prestigious industry events. Looking these influencer blogs, it is rather easy to cast a harsh critical lens on the influencer econ-omy, considering that it commoditizes individuality and encourages the performativity of everyday life.

In many respects, the popular perception of influencers is that they engage in a lot of online hustling, posturing and self-mythologizing – while wielding the self as a saleable object. Indeed, Henry Giroux's misgivings about the increasing dominance of neoliberal techno-capitalism are apropos here – though Hong and many of her peers do speak openly in support of various issues of social urgency, from destructive wildfires and ecological crises, to labour exploitation, as well as expressing misgivings around their complicity with a rapacious capitalist economy. In 2019, Hong launched a podcast called *Vanessa Wants to Know* in which she interviews influential figures in the worlds of fashion and media: a move that gestures towards greater substance while also maintaining the integrity of her brand. Unlike many of her contemporaries, Hong has attempted to directly address her racial identity, dedicating the first season of the podcast to spotlighting Asian excellence:

> We launched season one of *Vanessa Wants to Know* (*VWTK*) earlier this year, which I dedicated to Asian excellence. I felt, for our opening season, it had to be a topic extremely important to me. Growing up first generation Chinese–Canadian, I never had an example of what an Asian in mainstream media, killing it, would look or sound like. *VWTK* gave me that opportunity. Many of the guests I spoke with are not only stars in their respective fields, but true visionaries—whether it be Phillip Lim redefining what it means to be a designer today in 2019, or Michelle Lee, editor-in-chief of *Allure* magazine, using beauty as a vehicle for deeper cultural conversations. (Hong 2019: n. 1)

While not politicizing her subjects, Hong does speak about the paucity of Asians in high fashion, especially in top positions, as well as the need for greater visibility, diversity and structural change in mainstream media. I acknowledge that most scholarly reflections on the influencer economy characterize it as the underbelly of global techno-capitalist superficiality and that very well may be true. Although, as stated, I'm more interested in the psychodynamics of self-imaging as it expresses both individual and collective desires for recognition.

SOMEONE WHO LOOKS LIKE ME

Charles Taylor's writings on the politics of recognition consider the importance of recognizing the contributions and cultural histories of underrepresented groups (Gutmann in Taylor 1992: 3). The argument for recognition claims that minority groups (including women, immigrants and gays) have been 'forced to adopt a depreciatory image of themselves [...] they have internalized a picture of their own inferiority', which creates a debilitating low self-esteem (Taylor 1992: 26). Taylor argues that White society has constructed and projected a demeaning picture of certain minority groups for so many generations, that they have internalized and adopted an attitude of crippling self-depreciation (Taylor 1992: 26). Contrarily, for Appiah, the notion of internalization of what the scholar calls a negative 'life-script' is problematic in Taylor's formulation because it does not consider the complexity of identities (or the myriad differences, values, political views, etc.) among minority groups. Appiah argues that demanding respect for minority groups requires that there are scripts that go with being racially Othered: that there are proper ways of being women, or Asian (for example) – expectations to be met, and demands made (Appiah in Taylor 1992: 162). This expectation functions to replace 'one kind of tyranny with another' (Appiah in Taylor 1992: 162). But perhaps more urgent is Appiah's notion that it is the recognition of racial difference by White society that requires specific 'scripts' to be followed, scripts that forever lock communities of colour in a confining set of stereotypical identifications (Appiah in Taylor 1992: 163). That certain communities must organize their lives around these stereotypical scripts of racial identification (as a means to achieve recognition) is crucial to our understanding of self-imaging in online cultures. Appiah makes a similar argument in his *New York Times* op-ed, which argues that the need to *see someone who looks like me*, projected on the screen: to have one's desire for recognition fulfilled, necessitates racial reduction and the ideological oversimplification of our identities.

The more cynical interpretations of the selfie/influencer debate share a central conceit that concerns me, which is not simply what I would characterize as a careless or callous disregard for the deeply personal motivations for self-imaging – but rather the disdain articulated around the private desires expressed in individual images. There is much more interest in viewing the selfie judgmentally as a social scourge of sorts, as a representational phenomenon defined by a society where private citizens don't feel they exist without photographic evidence. In fact, the negation of individual desire extends beyond this very contemporaneous discussion. And yet, we must come to terms with this insistent need to characterize desire as a disruption or as a regressive tendency that fosters 'philosophical myopia, encouraging one to only see what one *wants*, and not what *is*' (Butler 1999: 3, original emphasis). Within this particular discussion, such attitudes take the shape of expressed misgivings about the rapaciousness of neoliberal capitalism while ignoring the very human longings and desires for recognition that emanate from these forms of representation.

Judith Butler, in the previous quotation, expresses a critical concern regarding the philosophical tendency to 'obliterate' desire through the formulation of 'strategies to silence or control it' (Butler 1999: 2).

The domestication of desire in the name of reason or, more aptly, the constructed philosophical split between reason and desire is the subject of Butler's ire – as it expresses a contradictory moralism. Desire is in fact the operation implicit to the act of self-imaging, which is its *modus operandi*. The envisioning of desire is not the antithesis of tacit knowledge, any more than there is an irrational desire. And as Butler argues, 'there is no necessarily irrational desire, no affective moment that must be renounced for its intrinsic arbitrariness' (Butler 1999: 2). That implicit arbitrariness of desire externalized is what intrigues me about online self-imaging: that tension between consciousness and self-consciousness that offers up the intimacies and longings of an individual. The photographic self-portrait is an intensely psychological image that mobilizes (or more aptly weaponizes) the returned gaze in the process of self-conception. In her characterization of portraiture as a profoundly psychological form of art, Cynthia Freeland suggests that within the genre, there are two fundamental yet conflicting aims: 'the revelatory aim of faithfulness to the subject, and the creative aim of artistic expression' (Freeland 2007: 97):

> The portrait encompasses distinct and even contradictory aims: to reveal the sitter's subjectivity or self-conception; and to exhibit the artist's skill, expressive ability, and to some extent, views on art. But historically this second aim was more restricted than we now imagine, and reciprocity was not the dominant paradigm for the painter/sitter relationship. (Freeland 2007: 97)

Freeland's intent is to illuminate the dualities of portraiture and to better understand its psychological dimensions. What is it about portraiture that is so compelling; from where does its gravitas issue – is it in its ability to reveal the essence of the sitter, or does it perhaps reside in the more technical and/or formal dimensions and characteristics that are unique to the image maker's gaze? Regarding the selfie, my interest lies in the self-portrait's ability, as Freeland articulates, to reveal the subjectivity (the essence) of the subject of representation: a characteristic that is more fraught and intellectually (and representationally) complicated in its self-obsessions. Does the selfie reveal essences about the subject or does it reveal something about the culture that produces it? For Freeland, the core tension in portraiture resides in the genre's demands on truth (the veracity of the photographic image) vs. the expressive vision of the image-maker. But how is this conflicting relation transformed when the image-maker is also the subject?

I would agree with Freeland that, above all things, the portrait is meant to convey the *person-ness* of the subject, which she characterizes as the central aim of modern conceptions of portraiture – that the image-maker:

> [s]eeks to convey the subject's unique essence, character thoughts and feelings, interior life, spiritual condition, individuality, personality, or emotional complexity. Just how this is done involves use of the varied techniques of portraiture to show many significant external aspects of a person, such as physiognomy, in addition to the depiction of features such as status and class through the use of props, clothing, pose and stance, composition and artistic style and medium. But ultimately we expect a good portrait to convey the person's subjectivity. The sitter should appear to be autonomous and a distinct person, with unique thoughts and emotions. As a person, the sitter is embodied, but the self is there 'in' the embodiment and the artist must 'realize', 'concretize' or 'objectify' it in the image. (Freeland 2007: 98)

The ease with which the selfie concretizes and objectifies perhaps taps into those aspects of self-portraiture that generate the most suspicion and ire – not least because of the image-maker's apparent inward turn, or the so-called narcissistic compulsion, where adulation becomes self-obsession. It is in this turning inward, in its more performative dimension, that the selfie (or self-portrait) deviates so dramatically from traditional portraiture. In this sense, the selfie – as a form of self-reflective image making – bears more in common with a lineage of performative self-portraiture within the history of art. Feminist art historian Amelia Jones, in her essay 'The Eternal Return: Self-Portrait Photography as a Technology of Embodiment', discusses a range of artists whose highly conceptual engagements with self-portraiture establishe an exaggerated mode of performative self-imaging that opens up an entirely new way of thinking about photography and the racially, sexually and gender-identified subject. In her engagement with performative self-portraiture in identity art practices, Jones discusses the work of several artists who utilized the photographic medium to create narrative-based and highly stylized images. In the process, they reimagined the genre of self-portraiture, as they wielded the self in an activist mode. Jones's critical engagement with identity-based forms of self-imaging suggests something rather meaningful to our ever-evolving understanding of the selfie: that these exaggerated forms of self-portraiture stage the body; that they in essence freeze the body '*as* representation and so – as absence, as always already dead –

in intimate relation to lack and loss' (Jones 2002: 949). The claim being made here is that the photograph is essentially a death-dealing apparatus in its ability to 'fetishize and congeal time' (Jones 2002: 949). This point is useful to our reading of online self-imaging, because selfies, in their exaggerated theatricality (to borrow Jones's language), highlight the fact that self-portraiture is inherently performative, and by extension, it has a tendency to rejuvenate or reanimate the genre. The selfie has productively restored the photographic self-portrait, particularly its ability to foreground the self, as well as the creative 'mobilization of technologies of representation, by performing the self through photographic means' (Jones 2002: 950).

In relation to Jones's assessment of performative self-portraiture in high art, the selfie foregrounds how identities become constituted in representation, in that online self-imaging thrusts otherness outward – projecting it into the foreground. To a large degree, a selfie does not only foreground the representational complexity of the self in relation to the other but also the self *as* other. Along these lines, Jones claims that the performance of the self is not 'self-sustaining or coherent in itself, not a pure, unidirectional show of individual agency, but always contingent on otherness' (Jones 2002: 971). In essence, the selfie objectifies the sitter, rendering the self as irrecoverably other: a spectacle. For that reason, the selfie is the perfect representational medium for visually challenging regulatory norms around race, gender, sexuality, class and national identity. And therefore, the staging of the body as cultural critique is perhaps the driving force of the most progressive and oppositional forms of self-imaging online. However, Jones's notion of the self-portrait as projecting difference or alterity outwards also gives some potential insight into the tendency towards judgment, pejorative insult and the apparently gendered pathologizing of the selfie taker as narcissistic, in that the spectacle of difference (in representation) is always read through the lens of incoherence and lack – and hence as a form political troubling. Bodily spectacle is all too often read reductively as alterity, and its fetishization, if perceived as excessive and fixated, is often read (at best) as a neurosis if not as entirely irrational and pathological.

In response to these less charitable characterizations of online self-imaging, I argue that it is not just desire (the desire of the Other) that unsettles, but rather the self-eroticizing and objectification: what is perceived as a narcissistic self-display that positions the female body as a fetish object. This articulated expression of female agency *is* scopophilic, but its voyeuristic pleasures are self-fixated – even while visually acknowledging (through a performed femininity) that the female body (in representation) is not her own, but a matter of public concern.

Female bodies are always subjected to the gaze and always rendered *the object* of popular representation. Even though the visual codes of female display are manipulated to grand effect, there is an implied pleasure in directing the gaze back upon itself, while playing with the 'to-be-looked-at-ness', that Laura Mulvey theorized in her famous essay on visual pleasure (Mulvey 1989: 14). Rather than reject the passive female/active male binary, implicit to the selfie, is a claimed pleasure – an active *jouissance* of sorts, that gestures towards a femininity that is beyond repression, despite its many glaring contradictions.

Lacan's *jouissance* (or the pleasure principle) functions as a limit of enjoyment. It is a repressive law meant to prohibit the subject from experiencing too much pleasure, but the consequence is that the subject is locked in an eternal struggle to transgress the imposed boundaries. The subject's attempt to transgress the pleasure principle does not lead to more pleasure, but pain itself; a type of suffering or *painful pleasure*. This suffering is what Lacan calls *jouissance*. In essence, *jouissance* represents a paradox, in that the subject becomes conflicted with his/her own pleasure and derives satisfaction in the suffering brought about by the pleasure experienced (Lacan 1981: 183).

Jacques Lacan first developed his concept of an opposition between *jouissance* and the pleasure principle (what Sigmund Freud calls the pursuit of enjoyment) in his seminar 'The Ethics of Psychoanalysis' (1959–60). Lacan argued there is a *jouissance* beyond the pleasure principle, which compels the subject to constantly attempt to transgress the prohibitions imposed on his/her enjoyment – to go beyond the pleasure principle. Yet, according to Lacan, the result of transgressing the pleasure principle is not more pleasure, but pain, since there is only a certain amount of pleasure that the subject can endure. Beyond this limit, pleasure becomes pain, and this painful principle is what Lacan calls *jouissance*. *Jouissance* is therefore suffering.

FEMALE *JOUISSANCE*, SELF-LOVE, DESIRE AND RECOGNITION

Lacan's model has become a powerful metaphor for the contradictory tensions between pleasure and guilt: a paradox that informs individual action but also larger social relations. It is precisely the opposition between enjoyment and pain, vis-à-vis the imaging and visual consumption of Otherness that is of interest to this investigation. What happens when difference (in this case, the female body) enters into the realm of representation, in a cultural climate where gender inequity is a persistent problem? As a key psychoanalytic concept in Lacan's building upon Sigmund Freud, the concept of *jouissance* is phallo-centric and rooted in the repression of masculine pleasure. However, as woman is configured as lack (as non-universal), there is what can be called a female *jouissance*, or more appropriately the *jouissance of the Other*. Lacanian feminism has been an extremely generative in terms of our application of this concept – not least because the French philosopher's writing was instrumental to the linguistic turn in post-structuralism of the 1980s. As Nina Lykke points out, 'French theorists became important sources of theoretical inspiration for the continued feminist struggle against biological determinism' (Lykke 2010: 97). She continues, 'Some feminists found Lacan's orientation toward language a particularly useful aspect of his reinterpretation of Freud's psycho-analysis' (Lykke 2010: 97).

French feminist writer Hélène Cixous's formulation of Lacan's *jouissance* is more concerned with a woman's pleasure: asserting that *jouissance* is the root of a woman's creative power, and its suppression restrains the female subject's expressive potential. Therefore, we might think of feminist *jouissance* as a kind of transcendent state that signifies freedom from societal repression. Scholar Jane Gallop offers a similar interpretation:

> When *jouissance* becomes an emblem of French feminine theory, however, it is specifically identified as non-phallic, beyond the phallus. But even though *jouissance* is specified as feminine, the tendency to stiffen into a strong, muscular image remains. The difference between *jouissance* and pleasure is generally under-stood to be one of degree: *jouissance* is stronger and so the person who experiences it is stronger, braver, less repressed, less scared. (Gallop 1983: 114)

> Fear also appears in conjunction with *jouissance* in an English translation of French feminism, but otherwise. The editors of New French Feminisms state, in a footnote, that '[*jouissance*] is a word used by Hélène Cixous to refer to that intense, raptur-ous pleasure which women know and men fear.' Here the two are con-joined but divorced: we have *jouissance*, they fear it. If *jouissance* is defined, as it is by Barthes and the women, as a loss of self, disruption of comfort, loss of control, it cannot simply be claimed as an ego-gratifying identity, but must also frighten those who "know" it. As *jouissance* becomes a banner and a badge for French feminine writing, the accompanying fear or unworthiness is projected outward and we-militant and bold-lose the ambig-uous link to fear and emotion, which are catapulted beyond the

jouissance principle where it might even be their momentary fate
to take up residence in that mediocre and unworthy word, 'pleasure'. (Gallop 1983: 114)

In response to Gallop and the critical framings of other notable Lacanian feminists, we can see how *jouissance* can be mobilized in the
service of dismantling gender-based repressions. *Jouissance* holds a
particular fascination when unpacking the role that difference plays
in the representational schemas of a society where gender hierarchy
is fundamental to the social, economic and political order. I find these
theorizations incredibly useful when contemplating the psychological implications of the photographic self-portrait. As Amelia Jones
asserts, there is 'something fundamental about the body in relation to
the image, something that, indeed, provided the major impetus to the
development of photographic technologies: the desire for the image to
render up the body *and thereby the self* in its fullness and truth' (Jones
2006: xiv). Jones makes a salient point about the technology of the
camera as a means to render the body via the manipulation of light:

> Being an indexical trace of the body before the camera, then, the
> photograph promised to return the represented body to some
> kind of authentic state. Because the photographic portrait docu-
> ments the embodied trace of the self (with the mind made visible
> only through its body-sign), it highlights both the inextricability
> of body and mind and the fact that we often access the self via
> its visible – corporeal – form, a form we want to serve as guar-
> antor of the body. The photographic portrait seems to reaffirm
> the body's never-ending 'thereness,' its refusal to disappear, its
> infinite capacity to render up the self in some incontrovertibly
> 'real' way. (Jones 2006: xiv–xv)

What Jones argues for here is a more nuanced engagement with our
relation to the representation of bodies – further suggesting that we
have an attachment to their presumed legibility: that there is a desire
for the body and image to function as a clear reflection and indicator for ideological meanings. In other words, we often interpret and
engage others through their appearance, which is the basis for stere-
otyping, racism and forms of social discrimination (Jones 2006: xvii).
Kwame Anthony Appiah, in his thoughts on recognition and popular
representation, suggests something similar. He argues that the need
for a representative image of ourselves requires certain reductions that
are ultimately less about seeing *someone who looks like me*, as they
are a dubious confirmation of one's existence. The types of self-images

circulating online (selfies and influencer portraits) 'deploy technologies of visual representation to render and/or confirm the self (paradoxically: objectifying the self so as to prove its existence as a subject), and the way in which these technologies expose the inexorable failure of representation to offer up the self as a coherent knowable entity' (Mulvey 1989: 19). The self-fashioning of desire implicit to the online self-image does fulfil that 'primordial wish for pleasurable looking', that Mulvey articulates, fully indulging the scopophilic and narcissistic pleasures of recognition (Mulvey 1989: 19).

The vast landscape of selfies and influencers is by no means strictly female, though I assert that the most urgent and progressive interventions online are driven by women. In fact, I have written before that within the female-driven landscape of 'selfie-laden' blogging and influencer culture, the spectacle of female bodies is the dominant driving force. But these images may constitute *a new love affair* (to Borrow Mulvey's terminology) between the image and the self-image (Mulvey 1989: 19). And, it is true that the visual power of online self-portraiture is rooted in a type of pleasure that is voraciously claimed: an oppositional desire and enjoyment in oneself as a response to a culture of devaluing and misrepresentation (Murray 2015: 560). This enjoyment is bound up in a conglomeration of intersecting effects around self-love, desire and the need for a mutual recognition: one in which both the subject of representation and the viewer/consumer enjoy the visual pleasures of the self as spectacle. This reciprocity is likely the driving force behind the selfie/influencer matrix (particularly in relation to its mobilization by women of colour): that it enables both the subject and the viewer/consumer to gaze lovingly upon that which has historically been denied – the imago of the devalued and the underrepresented, presented within the glitzy motifs and vacuity of consumer techno-capitalism. But despite its existential flaws, it *is* a new love affair. And perhaps in the glamorous world of online self-imaging, that self-love is the greatest love of all.

References

Appiah, Kwame, Anthony (1994), 'Identity, authenticity, survival: Multicultural societies and reproduction', in A. K. Appiah (ed.), *Multiculturalism: Examining the Politics of Recognition*, New Jersey: Princeton University Press, pp. 149–61.

Appiah, Kwame Anthony (2019), 'What does it mean to "look like me"?', *New York Times*, 21 September, https://www.nytimes.com/2019/09/21/opinion/sunday/minorities-representation-culture.html?searchResultPosition=17. Accessed 12 January 2020.

Butler, Judith (1999), 'Introduction', *Subjects of Desire: Hegelian Reflections in Twentieth- Century France*, New York: Columbia University Press, pp. 1–16.

Chun, Wendy Hui Kyong, (2012), 'Race and/as technology, or how to do things to race', in L. Nakamura and P. Chow-White (eds), *Race after the Internet*, New York: Routledge, pp. 38–60.

Freeland, Cynthia (2007), 'Portraits in painting and photography', *Philosophical Studies: An International Journal for Philosophy in the Analytic Tradition*, 135:1, pp. 95–109.

Gallop, Jane (1984), 'Beyond the *jouissance* principal', *Representations*, 7, Summer, pp. 110–15.

Gamson, Joshua (2011), 'The unwatched life is not worth living: the elevation of the ordinary in celebrity culture', *Theories and Methodologies*, 126:4, pp. 1061–69.

Giroux, Henry (2015), 'Selfie in the age of corporate and state surveillance', *Third Text*, 29:3, pp. 155–64.

Gómez, Edgar and Thornham, Ellen (2015), 'Selfies beyond self-representation: the (theoretical) f(r)ictions of a practice', *Journal of Aesthetics & Culture*, 7:1, pp. 1–10.

Gutmann, Amy (ed.) (1992), 'Introduction', in C. Taylor (ed), *Multiculturalism and the Politics of Recognition*, New Jersey: Princeton University Press, pp. 1–3.

Hong, Vanessa (2019), 'A moment with Vanessa Hong', *Live the Process*, 18 December, https://livetheprocess.com/blogs/transformational/a-moment-with-vanessa-hong. Accessed 3 February 2020.

Jones, Amelia (2002), 'The eternal return: self-portrait photography as a technology of embodiment', *Signs*, 27:4, pp. 947–78.

Jones, Amelia (2006), 'Prologue', *Self/Image: Technology, Representation, and the Contemporary Subject*, London: Routledge, xiii–xix.

Khamis, Susie, Ang, Lawrence and Welling, Raymond (2017), 'Self-branding, "micro-celebrity" and the rise of social media influencers', *Celebrity Studies*, 8:2, pp. 191–208.

Lacan, Jacques (1981), *The Four Fundamental Concepts of Psychoanalysis*, New York: W.W. Norton & Company.

Lykke, Nina (2010), *Feminist Studies: A Guide to Intersectional Theory, Methodology and Writing*, London: Routledge.

Mulvey, Laura (1989), 'Visual pleasure and narrative cinema', *Visual and Other Pleasures*, Bloomington: Indiana University Press, pp. 14–38.

Murray, Derek Conrad (2015), 'Notes to self: The visual culture of selfies in the age of social media', *Consumption Markets & Culture*, 18, pp. 490–560.

Murray, Derek Conrad (2018), 'Selfie consumerism in a narcissistic age', *Consumption Markets & Culture*, 23:1, pp. 21–43.

Murray, Derek Conrad (2020), *Visual Culture Approaches to the Selfie*, New York and London: Routledge.

Ng, Konrad (2016), 'Online Asian American popular culture, digitization, and museums', in S. Davé, L. Nishime and T. Oren (eds), *Global Asian American Popular Culture*, New York: New York University Press, pp. 139–50.

Taylor, Charles (ed.) (1992), *Multiculturalism and the Politics of Recognition*, New Jersey: Princeton University Press.

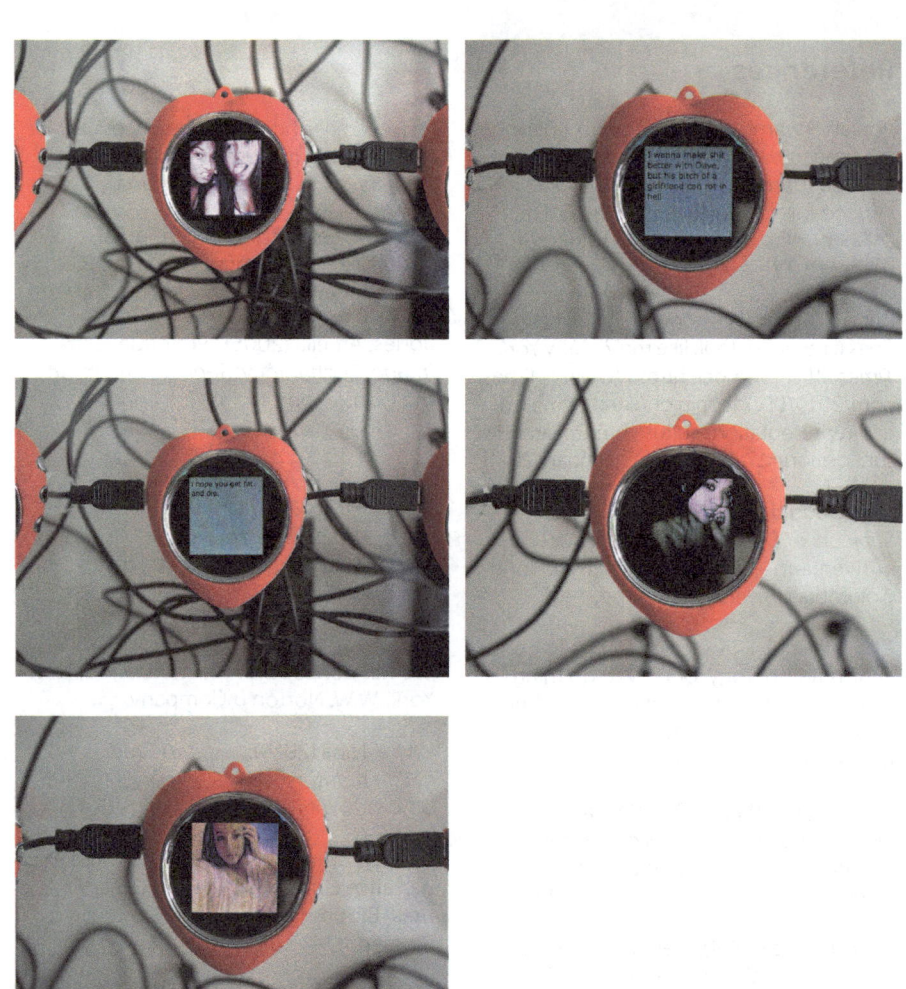

FIGURE 5.3–5.7: Jeroen van Loon, *Kill Your Darlings*, 2012. Installation, 97 LCD displays, 10 USB hubs, wood, plexiglass, 120 cm x 120 cm x 18.8 cm. Courtesy of the artist.

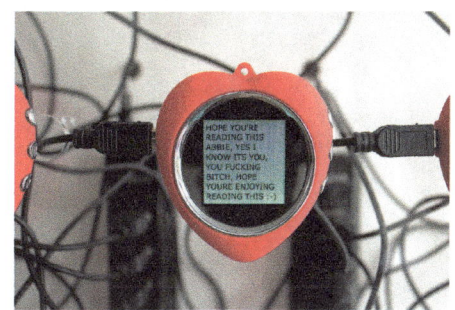

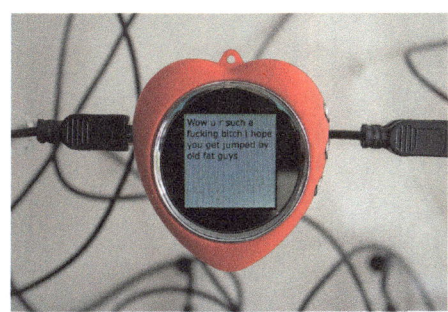

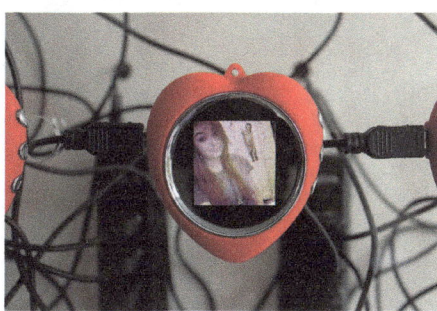

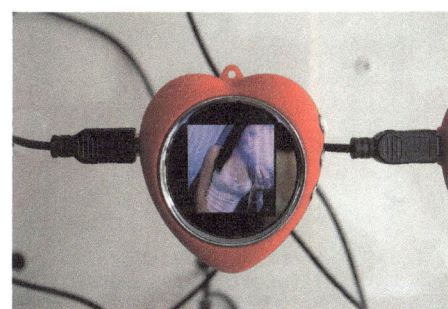

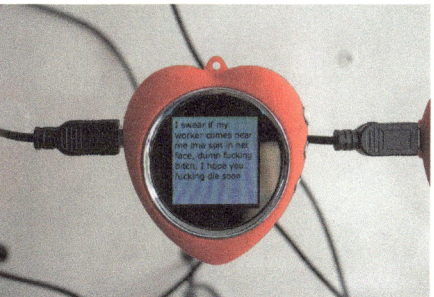

FIGURE 5.8–5.12: Jeroen van Loon, *Kill Your Darlings*, 2012. Installation, 97 LCD displays, 10 USB hubs, wood, plexiglass, 120 cm x 120 cm x 18.8 cm. Courtesy of the artist.

6
Digital Onscenities

'I know it when I see it', said US Supreme Court Judge Potter Stewart in 1964 on the elusive definition of hard-core pornography. Despite such evident outlooks, the perspectives on the obscene – its reception and perception take different forms depending on the modes and vehicles of representation. Before the era of analogue media (e.g. photography or television), watching for erotic pleasure was more private, and confound to immediate – even if non-contact interactions with a naked body. Despite that physical immediacy, the 'old-fashioned' In Real Life (IRL) analogue voyeurism was very restricted, mostly illegal and controlled at the limits of the law, depending on the political systems and times – which made it a marginal, secret and morally reprehensible activity. The paradigm shifted with the advent of digital voyeurism in the twenty-first century, when it became widespread in all spheres of society, legitimized and institutionalized.

Today, a self-confident voyeurism – explicitly and implicitly pornographic – is the new norm. So is the self-confident exhibitionism – consented, chosen staged for the voyeur who has become legitimate. Watching, following, peeping, as a standard of social interaction, make the obscene leave its traditional senses, turning the marginal pleasure into popular entertainment.

Peeping Tom (Porn Version) is a video installation by Thomas Israël that engages with the new status of porn and voyeurism by allegorizing the compulsion of watching and being watched. The large eyeball, disintegrated from the rest of the body, follows the audiences as they pass by, leaving them with the oppressive sensation of being spied on or even stalked. The installation singles out the eye as the major organ of excitement to speak of the mediated, doubled vision possible due to the camera lenses, or the lenses of media forms, their interfaces, formats and surveillance options. The flesh offered in the artwork is both fascinating and disgusting. The effect emerges from a confrontation with the technological disintegration and degradation of the body, which even if does not disintegrate and degrade the voyeuristic pleasure in us, disintegrates and degrades our intimate experience. Exposed to the presence of the 'naked eye', the audiences become subjected to what they subject others; in this way, they get confronted with the effects of their own scopophilia.

Peeping Tom (Porn Version)

THOMAS ISRAËL

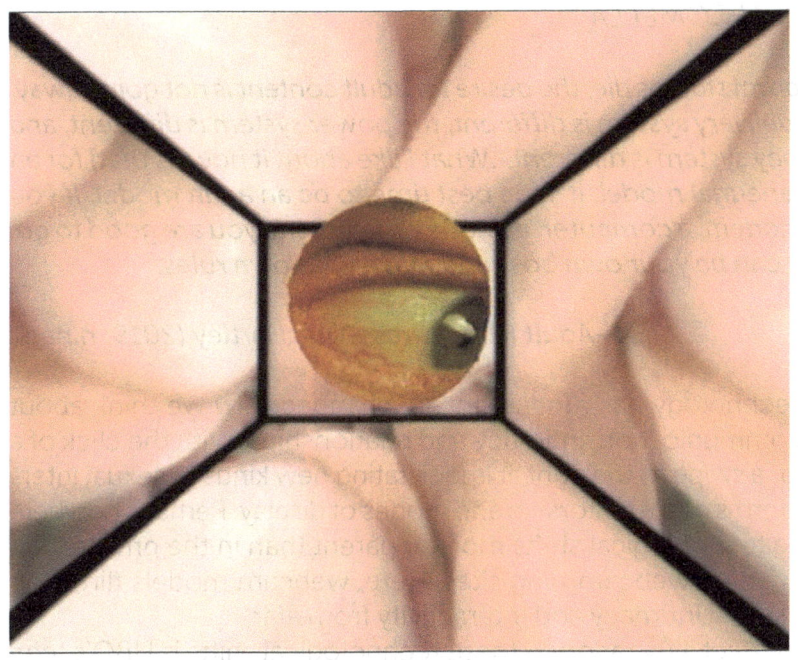

FIGURE 6.1: Thomas Israël, *Peeping Tom (Porn Version)*, 2006. Interactive installation, computer, video projector or screen, camera, specific software. Courtesy of the artist and Galerie Charlot, Paris/Tel Aviv.

The New Onscenity: Navigating Digital Desires in the Twenty-First-Century Pornoscape

LYNN COMELLA

As the [porn] studios die, the desire for adult content is not going away, but the delivery system is different, the power system is different, and the money system is different...What I like about it now [is that] for an entrepreneurial model, it's the best time to be an adult model. If you have a modem, a computer, a phone or a camera, you are good to go, and you can be your own boss and make your own rules.

—Adult Film Actress Nina Hartley (2019: n.pag.)

Digital technology has fundamentally changed how we think about sexual communication, intimacy and relationships. With the click of a mouse or a swipe to the right, it is facilitating new kinds of sexual inter-actions, forms of spectatorship and modes of display. Perhaps nowhere are these technological shifts more apparent than in the private chat rooms of adult webcamming sites. Here, webcam models flirt, strip and sell sexual intimacy and interactivity from afar.

A perfect illustration of these technological shifts is HBO's teen drama *Euphoria*. Described as a 'pornographic after-school special' (Ahsan 2019), the show captures the digital zeitgeist of Generation Z narrated by 17-year-old Rue, an addict fresh from a stint in rehab with no intention of staying clean, viewers are plunged into the compli-cated dating lives and digital dramas of a sexually precocious group of Gen Z high schoolers. It is a world awash in selfies, sexting, dating apps and dick pics, one in which, according to Rue, sending nudes is 'a currency of love'.

Euphoria is not billed as a show about sex and technology and the sexualization of culture, but it very much is. Sex and digital culture permeate the lives of these teens. They fall asleep texting their sweet-hearts and wake up reaching for their phones. They have perfected the art of taking selfies and seek out the best light in their bedrooms for artistically rendered nudes. They navigate dating apps with ease. Their sexual desires and practices, moreover, are deeply informed by internet pornography as they strive to mimic the sexual poses, pouts and positions of the adult performers they have seen online. In a world sorely lacking in comprehensive sex education, pornography, for better or worse, offers them a visual language of sex.

Perhaps no character better illustrates just how complicated and contradictory these digital spaces can be than Kat. A feisty and prolific fan-fic author, Kat is determined to not start the school year as a 'prude', so she engages in a consensual sexual encounter at a party with a boy from a nearby school. When she learns that video of her having sex has been shared online, she panics. She forces a male friend to help take it down, threatening to go to the police and report that he is distributing child pornography. But then something happens. Kat sees that her video has been viewed online more than 85,000 times. Internet strangers love her, and a lightbulb goes off. Donning a black cat mask to shield her identity, Kat starts camming (despite the fact she is 16 years old and legally should not be) and is soon offering individual Skype sessions for hundreds of dollars to men who are willing to pay for her attention. Flush with cash, she changes her look, her wardrobe, her mannerisms and becomes, at least for a while, a more confident, sexually empowered version of herself. Kat has flipped the script and seized the means of producing her own sexualized image. No longer an unwitting object of nonconsensual digital voyeurism, she becomes a shrewd internet entrepreneur who trades her time and attention for online adoration and crypto currency.

Given *Euphoria's* no holds barred portrayal of teenage sexuality, dating, webcamming and pornography, it is perhaps not surprising that the National Center on Sexual Exploitation (formerly Morality in Media) accused HBO of promoting and normalizing sexually exploitive themes (2019). It is a charge that aligns with broader concerns regarding pornography and its allegedly harmful effects encapsulated in words such as 'pornified', 'pornification' and 'pornographication'. According to writer Pamela Paul, '[t]he pornification of American culture is not only reshaping entertainment, advertising, fashion, and popular culture, but it is fundamentally changing the lives of more Americans, in more ways, than ever before' (2005: 11). As scholar Clarissa Smith notes, however, terms such as 'pornification' have no precise meaning, and their use, which often comes from a place of alarm, obscures 'the specific histories and politics of both the cultural artefacts under examination and those who are doing the examination' (2010: 104).

Euphoria is an instructive point of departure for thinking about the growing availability of sex-tech and pornography not as a moral lesson or cautionary tale, but as part of larger economic, technological and cultural shifts that have resulted in new kinds of spaces and opportunities for digital intimacy, voyeurism and agency in the twenty-first century pornoscape. In describing these shifts, I use the language of 'sexualization' to explore the processes by which new forms of sexual explicitness appear on/scene. I begin by briefly discussing how, by the early 1970s, sex was becoming increasingly commercialized and accessible. From 1970s porno chic to the adult VHS revolution of the 1980s, on/scenity was prompting new kinds of cultural conversations and debates about the place of sex in everyday life, as well as producing new viewing experiences, both in public and the home. I then zoom in, first on adult entertainment trade shows, and then on the sphere of live adult webcamming, to illustrate how sex-tech in the digital era is creating new kinds of online spaces for sexual entrepreneurship, intimacy and voyeurism.

ON/SCENITY AND THE SEXUALIZATION OF CULTURE

In their sweeping history of sexuality in America, John D'Emilio and Estelle B. Freedman discuss the various factors that helped redefine post-war sexual mores, attitudes and behaviours. A growing consumer culture and the loosening of obscenity laws in the 1950s, coupled with the emergence of second-wave feminism and the gay and lesbian rights movement in the 1960s, 'initiated a reshaping of the nation's understanding of sexuality' (1997: 314). By the late 1960s, the belief that sexual pleasure was a legitimate and essential part of people's lives was widely accepted. From *Playboy* magazine to the birth control pill, sex was becoming big business and entrepreneurs were finding new ways to weave sexuality 'into the fabric of public life' (330) through advertising, entertainment media, advice columns and more (D'Emilio and Freedman 1997: 330).

What was previously off/scene was becoming increasingly 'on/scene'. The concept of on/scenity refers to the process by which material that was once considered taboo and obscene is brought into public view. According to film scholar Linda Williams, who coined the term, on/scenity is 'the gesture by which a culture brings on to the public scene the very organs, acts, "bodies and pleasures" that have heretofore been designated on–off–scene, that is, as needing to be kept out of view, locked up in what Walter Kendrick has named the Secret Museum' (1999: 282). On/scenity is a flashpoint for controversy and debate prompted by new forms of sexual explicitness that challenge taken-for-granted boundaries between public and private, prurient and ordinary, respectable and vulgar, categories that are also deeply shaped by gender, race and social class.

Debates regarding on/scenity and the sexualization of culture are nothing new and emerged with particular vigour in the United States during the era of late nineteenth and early twentieth-century

Comstockery. During this period, anti-vice crusaders, led by the zealotry of Anthony Comstock, waged war on everything from dime store novels to pornography to information leaflets about birth control. Central to the ideology of Comstockery was the belief that sexuality belonged in the private sphere, and any public expression of it was, by default, obscene (Friedman 2000; Strub 2010).

The rise of 'porno chic' in the 1970s thus gave new voice to old concerns. The 1972 release of *Deep Throat* was for many feminists a catalyzing moment that captured the 'painful truth' about how men really felt about women (Bronstein 2011: 82). Made on a shoestring budget of $25,000 by director Gerard Damiano, *Deep Throat* would go on to earn an estimated $100,000,000 worldwide. Starring adult film actress Linda Lovelace, the movie told the story of a woman whose lack of sexual response prompts her to visit a doctor for help. The doctor, played by actor Harry Reams, discovers that Lovelace's clitoris has mysteriously migrated to a location deep inside her throat. If she wants to experience what an orgasm feels like, the doctor tells her, she will need to learn how to perform 'deep throat' fellatio.

Deep Throat was a cultural phenomenon. Middle-class adults of all ages flocked to theatres to see it, as did Hollywood celebrities. The film was reviewed in the *New York Times*, and Lovelace herself was treated as a celebrity, walking the red carpet at events, landing on the covers of *Time* and *Newsweek* and appearing as a guest on the *Tonight Show* with Jonny Carson.

Gone, it seemed, was the social shame and taboo once associated with walking into an adult theatre to see a pornographic movie. Pornography was suddenly chic and on/scene, and efforts to ban the film from screening in certain cities only fuelled its popularity, piquing the curiosity of people who wanted to see for themselves what all the fuss was about. The film's long theatrical run, combined with its box office receipts, 'put to rest the idea that only a fringe audience of perverts and sexual deviants frequented porn films. To the contrary, middle-class Americans made up the largest and most rapidly growing segment of the market' (Bronstein 2011: 75).

It was against this backdrop that opposition to pornography began to grow. Women across the country denounced *Deep Throat* as sexist. It was not, they argued, a story about a sexually liberated woman, but rather it was just one more film that privileged male desire and the male gaze at the expense of women. Although it would be several more years until an organized feminist movement dedicated to fighting pornography would emerge, for many women, *Deep Throat* signalled a call to arms.

Central to debates about on/scenity is the role of technology. Sex has always been at the vanguard of technological innovation (Attwood 2010). As sex-tech entrepreneur Andrea Barrica notes, '[n]ew technology begins as the domain of experts and professionals – but the ability to access sexually explicit material is what brings in the masses' (2019: 97). Porn created a demand for internet speed and peer-to-peer sharing and propelled the development of online payment systems. Technological shifts also, perhaps inevitably, bring fresh critiques about the ways in which explicit sexuality is creeping into our everyday lives. Recall, for example, the 1995 *TIME* magazine cover story on the risks of cyberporn that featured an image of a wide-eyed child sitting in front of a computer keyboard, his mouth agape. The message was clear: cyberporn had arrived, and it was coming for your children.

Over the years, technological developments created new delivery systems for explicit sexual content, as well as opportunities for content producers to woo new viewing audiences that might be curious about catching a peek at what was previously off/scene. In the book *Smutty Little Movies*, media historian Peter Alilunas argues that the VHS revolution of the 1980s was a game changer for the pornography industry in more ways than one. Home video 'all but decimated the traditional adult theatre circuit, permanently changed the industry, and altered the cultural landscape' (2016a: 6). As home video became dominant, opportunities for previously marginalized producers and audiences soon followed. The combination of privacy, availability and technology, according to Alilunas, 'created the possibility to rethink the presentation of sexuality that, in some cases, embraced an unabashed celebration of women's pleasure' (2016a: 117). The affordability of VHS technology meant that anyone could pick up a video camera and make the kind of porn they wanted to see, creating new markets not only for straight women, but couples, lesbians, bisexuals and queer audiences (see Queen 2015; Rednour and Strano 2015; Strub 2015). These changes also resulted in the creation of new distribution streams and spaces for consumption. From feminist sex-toy stores to suburban homes, VHS helped democratize access to pornography and the viewing pleasures that came with it (see Juffer 1998; Comella 2013, 2017).

It was not that long ago when rows and rows of adult videos and DVDs dominated the shelves of adult bookstores, an image that seems almost quaint by today's standards of streaming video, adult webcamming and porn clip sites, such as ManyVids, that offer custom content and one-to-one interactions between fans and performers via texting, video chats and phone calls. The business of pornography has shifted dramatically in just several decades. One place that provides a consistent look at emerging sex-tech trends and their reception are adult industry trade shows.

TAKING THE PULSE OF THE ADULT INDUSTRY

Every January the Adult Video News (AVN) Adult Entertainment Expo sets up shop in Las Vegas. Described as the world's largest adult industry extravaganza, the four-day event, which culminates in the glitzy AVN Awards Show – affectionately known as the Oscars of Porn – attracts an estimated 35,000 people annually. The show represents different things to different people. It is at once a trade show, with business seminars and networking opportunities; a media and promotional showcase for new products, upstart companies and the latest technological innovations and an over-the-top, immersive fan experience in which performers in mesh body suits and barely-there-exotic dancewear canoodle, twerk, pose for photos and sign autographs as a way to say 'thank you' to the fans who not only make them into stars, but who also help keep the adult industry alive. It is also an opportunity for academic researchers to take the pulse of the industry, make research contacts with directors and performers and get a sense of what is trending commercially and what is not (see Comella 2010, 2014).

The Expo brings together under one roof the people, products and commercial interests that define a constantly evolving industry. In doing so, it makes visible what is often hidden from view: marketing and promotional apparatuses, performances of erotic labour and displays of porn fandom, laying bare the mechanisms that sustain a multi-billion-dollar industry that trades in commercially manufactured sexual fantasies and desires.

The show also functions as a cultural touchstone, a cliff notes version of an industry that remains somewhat enigmatic to the casual observer despite its cultural reach and influence. Celebrated writer David Foster Wallace shined a spotlight on the AVN Expo and Awards show with his longform essay 'Big Red Son' in which he sounds both enthralled and repulsed by the hyper-sexualized world he finds himself in and the people he meets (2005). The film *Inside Deep Throat*, the 2005 documentary about the making of the movie *Deep Throat* and the rise of 1970s porno chic, ends with scenes from the trade show floor juxtaposed with talking heads lamenting the rise of mass produced VHS pornography and the demise of porn's 'Golden Era' of the 1970s and early 1980s. Author and anti-pornography activist Gail Dines, for her part, has written about what she sees as the 'predatory capitalists' who come to the show promising empowerment but delivering subjugation instead (2011).

Organized by parent company AVN Media Network, the Expo is part of a larger adult industry apparatus aimed at both promoting and legitimizing a historically stigmatized industry. AVN was founded in 1982 when Paul Fishbein and two friends started a monthly newsletter in Drexel Hill, Pennsylvania geared towards educating video store buyers and consumers about the emerging adult video market. Created on a shoestring budget of $900, the first issue, which debuted in February 1983, was eight pages and had 20 subscribers (Alilunas 2016b). Since then, AVN has become an industry stalwart, with holdings that include several print publications, a digital publishing presence and the annual AVN Expo and Awards Show.

For many years, the Expo existed as an offshoot of the annual Consumer Electronics Show (CES), which also takes place in Las Vegas. Over time, members of the adult industry felt marginalized by the CES. As Foster Wallace notes, the CES treated the adult trade show as a kind of 'crazy relative in the family' (2005: 10), keeping it sequestered in an event space that was a bus ride away from other CES venues. According to AVN founder Fishbein, '[CES] took our money but didn't promote the adult event and the adult vendors were all over at the Sahara [instead of the Sands, where the CES was located]. Some companies came to us and asked us to do a stand-alone event' (2008: n.pag.).

In the late 1990s, the Expo became its own event that coincided with, but was organized separately from, the CES. In 2012, the show relocated to the Hard Rock Hotel and Casino, becoming a smaller version of its former self and aligning its brand with the Hard Rock's edgy vibe. It was a new era for the Expo, one marked by the marriage of sex and rock and roll, complete with AVN-branded promotional imagery spilling out from the convention hall floor and into the Hard Rock's restrooms and elevators and onto the casino floor.

Long gone are the towering mega-booths that once represented industry heavyweights such as Hustler and Vivid, complete with porn stars elevated on risers several feet in the air. In their place are adult webcam companies, such as MyFreeCams and Chaturbate, which are now not only the show's major sponsors, but occupy the prime real estate on the trade show floor.

As the organization of the pornography industry has changed, so too have modes of promotion, display and personal branding. Now, webcam performers, of various ages, body types, races and ethnicities, stand shoulder-to-shoulder behind tables branded with bold company logos, playfully flirting and hamming it up in front of laptop computers not only for fans at the Expo, but for paying clients in chatrooms who could be logged on, ostensibly, from anywhere in the world. Here, forms of erotic labour that are typically performed in more private settings are on full display.

The ways in which the AVN show and the wider adult industry have changed over the years are not lost on longtime performers who have had to navigate and adapt to new economic realities. As Wicked contract performer and 20-year industry veteran Jessica Drake notes, 'When piracy took over our industry around 2008, everybody had to change everything in order to survive and many did not. Or they did change, and they still didn't survive' (2019). When Drake entered the business, in 1999, it was still very star-oriented. Performers – many of whom looked like extras from *Baywatch* – aspired to be the biggest names they could and become contract performers with major studios. Adult studios such as Wicked, Vivid and VCA were churning out big budget films. A decade later, in the wake of the economic downturn and rampant piracy, a perfect storm was brewing. Adult DVD sales had slumped, and the industry was scrambling to stop the bleeding. Many companies were closing up shop or figuring out how to recalibrate their businesses with an eye towards survival. According to Drake:

> Everything was affected, from performer rates dropping because companies were trying to be more competitive [to other things]. We used to be paid to shoot a box cover separately from a movie. Being chosen as a box cover girl for a feature was...great. It was another job. You'd be paid $1000 to be on a box cover to promote this movie. That doesn't happen anymore. And magazine work used to be a thing. It's barely a thing anymore. (2019: n.pag.)

Although many adult industry professionals miss the studio era, and the money and star power that came with it, others have welcomed the new digital platforms and business models and the increased performer agency and autonomy they afford. But how, exactly, does adult webcamming work, and what is its appeal for both performers and consumers?

THE RISE OF ADULT WEBCAMMING

The phenomenon of webcamming is not new. In 1996, college student Jennifer Ringley turned on JenniCam and began documenting her daily life from her dorm room, foretelling the culture's interest in 'reality-based voyeurtainment' (Hart 2010) and helping to democratize 'exhibitionism for profit for the masses' (Yagielowicz 2015: n.pag.). Two decades later, by 2016, adult webcamming had solidified its status as an industry powerbroker. A January 2016 cover story in *AVN* magazine, for example, led with an article that asked whether cam girls were the new porn stars (Miller 2016). Several months later, *XBiz World*, another industry trade magazine, published a special issue on live adult webcamming, with an editorial describing it as 'the golden child of the online adult industry' (Parret 2016: 8). There are now adult webcam conferences, awards shows, webcam coaches and business seminars where webcam models – many of whom are women, but certainly not all – can learn how to build both a brand and a fan base and drive traffic to their chat rooms and money into their bank accounts. An estimated multi-billion dollar business that is said to account for more than one third of adult entertainment revenues globally, webcamming has been an industry game changer (Yagielowicz 2015).

New York Times reporter Matt Richtel has described the webcam industry as a 'kind of digital era peep show' (2013: n.pag.). According to sociologist Angela Jones, webcam models are a 'cohort of sex workers who use highly stylized chat rooms to sell a range of erotic fantasy to online voyeuristic patrons' (2015: 777). Similarly, anthropologist Sophie Pezzutto describes webcamming as an 'online strip show', in which cam performers use 'the webcam on their computer to put on a show for anyone in their chat room. The performer usually sets tipping goals and the more people tip by pledging tokens, the more happens on screen' (2018: n.pag.).

Cam models, who typically work from home or from established webcam studios that provide them with a private room, computer, webcam and high-speed internet, create public profiles and have a public chat room that might be hosted by any number of streaming sites, such as Chaturbate, Streammate or MyFreeCams. Models use their computers to project video of themselves to customers, who can use the default free 'chat' to communicate with models through typed conversations. Customers can visit a chat room as a guest and not pay anything, but they can also purchase tokens that they can use to 'tip' models for doing a dance or performing a strip tease, or to take a model into what Jones describes as a 'virtual private room' (2016: 229) for a more personalized interaction that might be sexual in nature but not necessarily. Models may perform alone or with someone else. They may dance, strip, masturbate or play with sex toys. It is not unusual for models to spend a significant amount of time simply talking to and getting to know the people who visit their chatrooms. As performer Betty Blac informs, 'I spend a lot of time in public chat and I would say that a lot of fans just want to talk to me. One person regularly comes, and he picks me just because he likes to find out what's going on in my life. We talk about everything under the sun' (Miller-Young 2015: 367). Blac's time 'on cam' might be spent talking about *Star Trek,* or answering questions about sex that a man visiting her chatroom might not feel comfortable asking anyone else. When she first started camming, she discovered that if she did something weird, like make balloon sculptures out of condoms she had blown up, men would tip her just because she was doing something they had never seen on cam before.

Jeze Bell, another webcam performer who cammed out of a studio in Las Vegas, similarly explained that although many of the men who visit her chatroom are looking for sexual gratification, others just want to hang out and chat. Sometimes she does naked yoga on cam, while other times she strips while hula hooping. When she first started webcamming, she had braces. 'They loved me for that. A guy would pay me just to put my hand in my mouth. His screen name was "tin grin"' (2015: n.pag.).

Unlike traditional porn content, which is directed by someone else, a webcam client is both a user and a director who, for a price, can tailor their interactions with a performer to fulfil a specific fantasy or desire: foot worship, small penis humiliation or financial domination. For the model, the exchange is a kind of temporary intimacy. 'It's real, but not real. They are real connections, but then you log off and turn off the computer', Jeze Bell explains. Different from working at a strip club or a legal brothel, the webcam performance is an interaction that requires no direct contact, creating a buffer of safety that appeals to many performers. Webcam clients, most but not all of whom are men, could be sitting in front of a computer anywhere in the world, thousands of miles away. They can see the model, but they cannot touch her – and she doesn't have to touch, or even see, them. According to Jeze Bell, 'You only do what you want to do on cam. You have control over your shows and control over your schedule. I don't know any other job that would grant me that freedom, flexibility, and money'.

Webcamming is a way for performers to bypass poorly paying minimum wage jobs and pay their way through school or support their families while also maintaining the kind of flexibility and autonomy they desire. For others, including more established porn performers, camming is part of what feminist labour scholar Heather Berg describes as 'porn's satellite industries', something they might do to supplement their porn work (2016: 161). Queer porn director and performer Courtney Trouble sees live camming as an integral part of what it means to be 'a sex worker in the digital age' (2019: n.pag.).

Although webcamming offers some of the same sexually explicit content found in pornography, the performance is live, not prerecorded, and highly interactive, rather than scripted, which 'allow workers and customers to create unique content for each performance' (Jones 2016: 229). The contact between a model and customer – which is mediated by a computer screen – makes the transaction different from the work of escorting or stripping. Adult webcamming defies easy categorization, and yet, at the same time, it shares a number of characteristics with other forms of sex work: it is a service, an erotic performance, a form of entertainment and an economic activity.

There are a number of factors driving the growth of the interactive webcam market. 'From the consumer standpoint', Stephen Yagielowicz from *XBiz* explains, 'live cams provide a high level of interactivity and personalization that simply cannot be matched by prerecorded photo/video content, while home internet connections have gotten speedy enough to where the quality of this experience is worth the expense… suddenly, the unattainable girl, guy, couple or other of your dreams is attainable 24/7'. Adult webcamming, he claimed, is 'reality TV at its best' (2016: n.pag.).

Adult webcamming has become a massive industry and according to one industry expert, its popularity has less to do with shifts in the porn industry and more to do with the demise of online dating and the migration of online relationships – and investor money – to adult webcam sites (Lee 2016). Men, it seems, are hip to the fact that many online dating sites are full of bots and fake profiles and have moved their business to adult cam sites, because there, they actually get the kind of interactions they are paying for. As Yagielowicz notes, 'there's a performer you can see and hear live and in real time, earnestly hoping to please you for profit' (2016). In addition, he adds, webcamming solves the problem of piracy and free porn, because the interactivity is so personalized it cannot be easily duplicated.

Researchers who study adult webcamming argue that camming is an example of feminist entrepreneurship that allows models to 'test their entrepreneurial ability' (Reece 2015: 270) by selling not only their online performances, but a lock of their hair, used panties and original custom content, as well as developing tech skills that are translatable to other jobs. Robert Reece suggests that camming's decentralized geography – it can be done anywhere there is a computer, camera and internet access – 'subverts the limitations of the traditional adult industry' (2015: 270), especially for those in marginalized communities, including cam models of colour and trans performers, who might not have access to major talent agencies or directors, but who can work independently and with ease from the safety and comfort from their homes.

The work of doing sex work is a growing focus for scholars studying the adult industry. Sophie Pezzutto, who has conducted extensive ethnographic research in Los Angeles and Las Vegas with transgender pornography performers, has written about the rise of what she calls the 'porntropreneur' (2019, 2020). According to Pezzutto, 'Performers today are better thought of as internet entrepreneurs, generating income from a range of activities beyond porn and using social media to market themselves' (2020: n.pag.). Even established porn performers now need side hustles to pay their bills, and they are turning to the adult 'gig economy' to earn income from multiple revenue streams that include, but are not limited to, camming, self-produced videos, selling subscriptions to online platforms such as OnlyFans and sexting with fans. Once upon a time, Pezzutto notes, porn stars were simply performers. Now, they are running small businesses that require a range of new skills. They have to be responsive to changes in algorithms and remuneration models and be technically savvy about operating various online platforms and apps. They have to be highly self-disciplined about scheduling their own productions and extremely mindful of their personal brands and online personas. Performers are also responsible for organizing their own photo shoots for various social media accounts, doing Q&As with fans on Instagram, posting behind-the-scenes content on Twitter and vlogging about their daily lives on YouTube. If it sounds exhausting, it is. In today's gig economy, there is no clocking in and clocking out, and no clear delineation about when the workday starts or ends. For performers, the demand to always be 'on' and provide fans with an intimate look into their everyday lives can be extremely labour intensive and emotionally taxing.

An important element of what fans are seeking is authenticity. Australian porn performer and legal scholar Zahra Stardust has examined the rise of pornographic authenticity, including the ways performers navigate demands for authenticity from both producers and fans (2019). According to Stardust, a focus on pornographic authenticity began to emerge in the 2000s, as feminist and queer porn producers worked to distinguish their films from the prescriptive formulas of mass-produced pornography. Authenticity requires performers to emphasize genuine pleasure and real orgasms and to 'share intimate and personal moments with audiences, both on film and social media' (2019: 2). But authenticity, Stardust notes, is also a performance, one that is carefully staged and curated to adhere to an aesthetics of ordinariness that itself has become a marketable commodity. Producing authenticity is, thus, another form of labour in a precarious adult gig economy in which performers are expected 'to foster relationships with consumers, to recruit and maintain membership bases for their websites, to attract clicks that can be converted into royalties, and to gain followers on social media in the hope of building an identifiable brand that can be easily searchable and booked for upcoming gigs' (Stardust 2019: 22).

Adult industry veteran Jessica Drake thinks authenticity is far more relevant today than it was in the early years of her career: 'Before, it was more about creating a very polished, finished image. That's great. Fantasy is fantastic, but at the same time, I have found that being very raw, really vulnerable and accessible, that creates more interest, more attraction' (2019: n.pag.).

Drake's first clue that fans were seeking greater access to unscripted moments of authenticity predated the rise of live webcam-ming and social media and was the result of an ordinary moment captured on film and included as a DVD extra. She had finished shoot-ing her sex scenes for the day, and was sitting on the floor, dressed in a t-shirt, eating a pie with a fork. According to Drake, she got more fan attention from those few seconds of unscripted ordinariness than she had for anything else up to that point in her career.

Now more than ever, porn performers and cam models are marketing and monetizing their authentic selves, rather than just sleekly polished or photoshopped versions that adhere to dominant ideals of female beauty and male fantasy. Authenticity is an increasingly essential part of building a personal brand and a social media follow-ing. By harnessing the aesthetics of ordinariness, performers are both subverting and democratizing ideas about who can be an adult star in the digital era. Realness – perhaps even more than fantasy – is now being packaged as a bankable commodity.

Catering to digital desires is undoubtedly big business; however, in the interconnected world of online media, performers' financial success is dependent on having unfettered access to social media outlets to grow their following, market their personal brands and direct fans to monetized content. The internet has become an especially fraught place for sex workers in the aftermath of the 2018 Stop Enabling Sex Traffickers Act (SESTA) and Allow States and Victims to Fight Online Sex Trafficking Act (FOSTA). Intended to fight trafficking, SESTA/FOSTA has actually made it more difficult for sex workers to operate safely online, because it holds website publishers responsible if third parties are found advertising prostitution on their platforms (Romano 2018). Following the law's passage, Craigslist ended its personals section and soon after, Tumblr decided to remove adult content from its platform altogether. As social media companies make changes to their terms of service agreements to limit sexual content, or worse, shut down sex worker accounts, performers' ability to take advantage of these platforms, and the global reach they enable, is increasingly tenuous.

Instagram, for example, deleted the account of trans performer Korra Del Rio just as she had reached 100,000 followers. Had she been able to convert just 3% of that following into customers willing to pay $10 for something — a video clip, a cam show or a text exchange — she would have earned $30,000, income that would have made a tremendous difference in her life (2019). The rise of internet censorship is severely affecting the ability of many sex workers to support themselves and exacting a psychological toll in the process.

CONCLUSION

New forms of sexual explicitness are increasingly on/scene and onscreen in the twenty-first century pornoscape. They are finding their way into our social media feeds and onto our dating apps, presenting new opportunities to see and be seen in the digital era. The world of pornography is perhaps more porous than ever, leading not only to the increased sexualization of everyday life, as captured by the teens on HBO's *Euphoria*, but to a fundamental realignment of power in an industry historically run by men for men. While the growth of sex-tech and the seemingly unbounded nature of pornography continue to be a sources of concern for parents and anti-pornography advocates, researchers are finding that using sex-tech can facilitate positive emotional and sexual connections, resulting in less loneliness and depression, not more. According to researcher Justin Garcia of the Kinsey Institute, academic studies are beginning to 'chip away at the idea that technology is replacing human connections and that love is becoming obsolete' (2019). The rise of sex-tech, including but not limited to the world of adult webcamming, highlights both the possibilities and challenges of navigating the complexities of digital voyeurism in safe and profitable ways, regardless of which side of the screen one is on.

References

Ahsan, Sadaf (2019), 'Why *Euphoria* feels like a pornographic after-school special with no goal other than to desensitize its audience', *National Post*, 27 June, https://nationalpost.com/entertainment/television/why-euphoria-feels-like-a-pornographic-after-school-special-with-no-goal-other-than-to-desensitize-its-audience. Accessed 20 February 2020.

Alilunas, Peter (2016a), *Smutty Little Movies: The Creation and Regulation of Adult Video*, Berkeley: University of California Press.

Alilunas, Peter (2016b), 'Bridging the gap: Adult video news and the "long 1970s"', in C. Bronstein and W. Strub (eds), *Porno Chic and the Sex Wars: American Sexual Representation in the 1970s*, Amherst and Boston: University of Massachusetts Press, pp. 303–26.

Attwood, Feona (ed.) (2010), *Porn.Com: Making Sense of Online Pornography*, New York: Peter Lang.

Barrica, Andrea (2019), *Sex-tech Revolution: The Future of Sexual Wellness*, n.p. : Lioncrest Publishing.

Berg, Heather (2016), '"A scene is just a marketing tool": Alternative income streams in porn's gig economy', *Porn Studies*, 3:2, pp. 160–74.

Bronstein, Carolyn (2011), *Battling Pornography: The American Feminist Anti-Pornography Movement, 1976–1986*, Cambridge: Cambridge University Press.

Comella, Lynn (2010), 'Remaking the sex industry: The adult expo as a microcosm', in R. Weitzer (ed.), *Sex for Sale: Prostitution, Pornography, and the Sex Industry*, New York: Routledge, pp. 285–306.

Comella, Lynn (2013), 'From text to context: Feminist porn and the making of a market', in T. Taormino, C. Parreñas

Comella, Lynn (2014), 'Studying porn cultures', *Porn Studies*, 1:1&2, pp. 64–70.

Comella, Lynn (2017), *Vibrator Nation: How Feminist Sex-Toy Stores Changed the Business of Pleasure*, Durham, NC: Duke University Press.

Del Rio, Korra (2019), interview by author, Henderson, NV, 9 July.

D'Emilio, John and Freedman, Estelle B. (1997), *Intimate Matters: A History of Sexuality in America*, Chicago: University of Chicago Press.

Dines, Gail (2011) 'Porn: A multibillion-dollar industry that renders all authentic desire plastic', *The Guardian*, 4 January, https://www.theguardian.com/commentisfree/2011/jan/04/pornography-big-business-influence-culture. Accessed 10 March 2020.

Drake, Jessica (2019), interview by author, Las Vegas, 23 January.

'Pilot' (2019), Augustine Frizzell (dir.), *Euphoria*, Season 1 Episode 1 (16 June, USA: HBO).

Fishbein, Paul (2008), email correspondence with author, 11 February.

Friedman, Andrea (2000), *Prurient Interests: Gender, Democracy, and Obscenity in New York: City, 1909-1945*, New York: Columbia University Press.

Hart, Hugh (2010), 'April 14, 1996: Jenni-Cam starts livecasting', *WIRED*, 14 April, https://www.wired.com/2010/04/0414jennicam-launches/. Accessed 20 February 2020.

Hartley, Nina (guest) (2019), 'AVN past and present with Nina Hartley, Lynn Comella and Miss Lollipop', *Peepshow Podcast*, 14 February, http://peepshowpodcast.com/peepshow-podcast-episode-42. Accessed 20 February 2019.

Jeze Bell (2015), interview by author, Las Vegas, 30 December.

Jones, Angela (2015), 'For black models scroll down: Webcam modeling and the racialization of erotic labor', *Sexuality & Culture*, 19, pp. 776–99.

Jones, Angela (2016), '"I get paid to have orgasms": adult webcam models' negotiation of pleasure and danger', *Signs. Journal of Women in Culture and Society*, 42:1, pp. 227–56.

Juffer, Jane (1998), *At Home with Pornography: Women, Sex, and Everyday Life*. New York: New York University Press.

Kinsey Institute (2019), 'Kinsey Institute "Sex-tech" Study Finds That Technology Facilitates Sexual and Emotional Interactions', https://kinseyinstitute.org/news-events/news/2019-11-21-sex-tech.php. Accessed 10 February 2020.

Lee, Ron (2016), telephone interview by author, 6 January.

Miller, Dan (2016), 'Are cam girls the new porn stars?', *AVN*, January, pp. 70–76.

Miller-Young, Mireille (2015), 'Race and the politics of agency in porn: A conversation with black BBW performer Betty Blac', in L. Comella and S. Tarrant (eds), *New Views on Pornography: Sexuality, Politics, and the Law*, Santa Barbara, CA: Praeger, pp. 359–70.

National Center on Sexual Exploitation (2019), 'HBO's *Euphoria* claims 'nudes are the currency of love' and promotes other sexually exploitive themes', https://endsexualexploitation.org/articles/hbos-euphoria-claims-nudes-are-the-currency-of-love-and-promotes-other-sexually-exploitive-themes/. Accessed 10 March 2020.

Parret, Don (2016), 'Editor's Note', *XBiz World*, November, p. 8.

Paul, Pamela (2005), *Pornified: How Pornography Is Transforming Our Lives, Our Relationships, and Our Families*, New York: Times Books.

Pezzutto, Sophie (2018), 'Why adult video stars rely on camming', *The Conversation*, 21 November, https://theconversation.com/why-adult-video-stars-rely-on-camming-104758. Accessed 10 March 2020.

Pezzutto, Sophie (2019), 'From porn performer to porntropreneur: Online entrepreneurship, social media branding, and selfhood in contemporary trans pornography', *AG: About Gender*, 8:16, pp. 30–60.

Pezzutto, Sophie (2020), 'The rise of the porntropreneur: Even Hustlers need side hustles in the gig economy', *The Conversation*, 21 January, https://theconversation.com/the-rise-of-the-porntropreneur-even-hustlers-need-side-hustles-in-the-gig-economy-129067. Accessed 10 March 2020.

Queen, Carol (2015), 'Good vibrations, women, and porn: A history', in L. Comella and S. Tarrant (eds), *New Views on Pornography: Sexuality, Politics, and the Law*, Santa Barbara, CA: Praeger, pp. 179–90.

Rednour, Shar and Strano, Jackie (2015), 'Steamy, hot, and political: Creating radical dyke porn', in L. Comella and S. Tarrant (eds), *New Views on Pornography: Sexuality, Politics, and the Law*, Santa Barbara, CA: Praeger, pp. 165–77.

Reece, Robert L. (2015), 'The plight of the black Belle Knox: Race and webcam modeling', *Porn Studies*, 2:2&3, pp. 269–71.

Richtel, Matt (2013), 'Intimacy on the web, with a crowd', *New York Times*, 21 September, https://www.nytimes.com/2013/09/22/technology/intimacy-on-the-web-with-a-crowd.html. Accessed 10 March 2020.

Romano, Aja (2018), 'A new law intended to curb sex trafficking threatens the future of the internet as we know it', *Vox*, 2 July, https://www.vox.com/culture/2018/4/13/17172762/fosta-sesta-backpage-230-internet-freedom. Accessed 19 March 2021.

Shimizu, C. Penley and M. Miller-Young (eds), *The Feminist Porn Book: The Politics of Producing Pleasure*, New York: Feminist Press, pp. 79–93.

Smith, Clarissa (2010), 'Pornographication: A discourse for all seasons', *International Journal of Media and Cultural Politics*, 6:1, pp. 103–8.

Stardust, Zahra (2019), 'From amateur aesthetics to intelligible orgasms: Pornographic authenticity and precarious labor in the gig economy', *AG: About Gender*, 8:16, pp. 1–29.

Strub, Whitney (2010), *Perversion for Profit: The Politics of Pornography and the Rise of the New Right*, New York: Columbia University Press.

Strub, Whitney (2015), 'Queer smut, queer rights', in L. Comella and S. Tarrant (eds), *New Views on Pornography: Sexuality, Politics, and the Law*, Santa Barbara, CA: Praeger, pp. 147–64.

Trouble, Courtney (2019), interview by author, Las Vegas, 26 January.

Wallace, David Foster (2005), *Consider the Lobster and Other Essays*, New York: Little, Brown.

Williams, Linda ([1989] 1999), *Hard Core: Power, Pleasure and the 'Frenzy of the Visible'*, Berkeley: University of California Press.

Williams, Linda (ed.) (2004), *Porn Studies*, Durham: Duke University Press.

Yagielowicz, Stephen (2015), 'The live adult webcam sector is thriving', *XBiz*, 30 June, https://www.xbiz.com/features/195717/the-live-adult-webcam-sector-is-thriving. Accessed 14 May 2021.

Yagielowicz, Stephen (2016), email communication with author, 8 January.

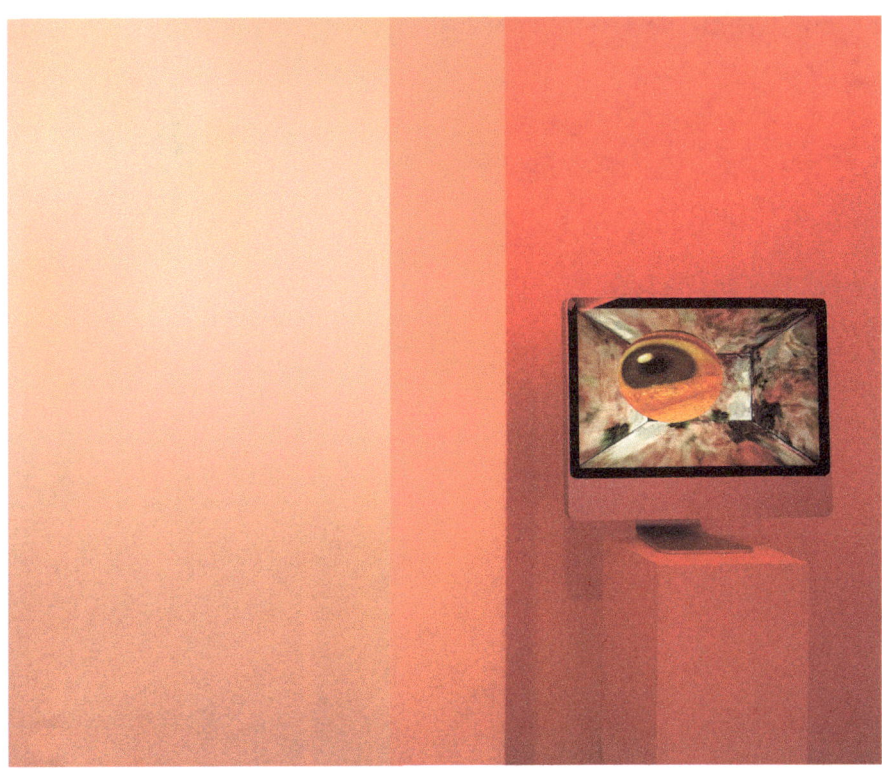

FIGURE 6.2: Thomas Israël, *Peeping Tom
(Porn Version)*, 2006. Interactive installation,
iMac, specific software. Courtesy of the
artist and Galerie Charlot, Paris-Tel Aviv.

7

Libidinal Techno-Scapes

Physical intimacy is something that happens in space and against different landscapes. Moving online, erotic practices have accommodated new backgrounds that change our erotic sexual behaviour, changing also our perception of sexual pleasure and pornographic imagery.

If asked about a difference between the Renaissance painting *The Birth of Venus* (1486) and a *Playboy* centrefold, most might say it is no contest: one is art and the other pornography. One is of human ideals, the other smut. But are Botticelli and Hugh Hefner really that different? Both project fantasy and erotic imagery through the media of their day. Both are vehicles of gender politics, defining standards of beauty and sexuality. What if adult performers – already mediated sex objects – struck 'classic' poses?

In *Webcam Venus* – a video art project from Addie Wagenknecht and Pablo Garcia – the artists asked online sexcam performers to replicate iconic works of art. This piece is an experimental homage to both fine art and the lowbrow internet phenomenon of cams. Sexcams use webcams and chat interfaces to connect amateur adult performers with an audience. Users log on to see men, women, transsexuals, couples and groups broadcast their bodies and sexuality live for the public, often performing for money. To create this experiment in high- and low-brow media, the artists assumed anonymous handles and spent a few hours each day for a month asking performers: 'Would you like to pose for me?'

By operating in the language of sexcams, the project alters the contemporary ideal of beauty with the ubiquitous display of sexuality online. It challenges the institutions that enforce false perceptions of propriety – via nudity in classical painting – as the only form of acceptable safe-for-work beauty. Publicly presented traditional paintings and sculptures are prevalent with sexuality and gender politics, and yet the display of nudity online is usually defined as 'pornography'. Amateur adult broadcasters also resist the popular, contemporary definition of beauty. They are not the typical definition of beauty prevalent in mainstream media: heavily photoshopped image in the name of advertising, which destroys self-image and confidence while encouraging materialism. Sexcam performers are the apotheosis of the most honest parts of us and yet, typically, the least valued part of a society. Even though they are transmitted virtually, they are real people and they are beautiful.

By researching interactive online spaces, the project was drawn to hubs where intimacy goes public – social media, blogs, webcams and chat rooms – and the idea they the content is accessible worldwide. The division between 'in real life' (IRL) and 'not in real life' (NIRL) is dissolving. Our relationships and most intimate interactions are no longer happening in the same room or even same language. With social media, developing a presence on the internet has become as simple as logging in. Opening your personal world to the outside world frames us each as our own brand – we maintain Twitter feeds and Facebook pages, promote our families and ourselves, we foursquare every place we go to and we Instagram everything we eat. We are becoming a society where we create, produce and consume all at the same time. In this paradigm of public intimacy, cybersex and sexcams not only seem less deviant, they practically seem inevitable.

Webcam Venus is also about networked cultures and digiphrenia: how technology lets us be in more than one space – or even more than one identity – at the same time. Sexcam performers construct identities through provocative handles, costumes, masks and interior decoration for the viewing audience. Their display of sexuality is part of this identity. When asked to pose in a 'classic' manner, sexcam performers become suddenly self-aware; they want to adjust their hair or surroundings to meet the request. For an instant, *Webcam Venus* reveals the identity that lives just outside the cam space; one where the person must improvize beyond the established protocols of adult performance. Different spaces and spatial contexts unveil different senses of pornography and erotic performance; we begin to see a paradox emerge: in real life, art nudes are acceptable, while naked bodies are inappropriate; in online life, graphic sex acts are acceptable yet de-sexualization on cam is difficult for some to maintain.

Webcam Venus

ADDIE WAGENKNECHT AND PABLO GARCIA

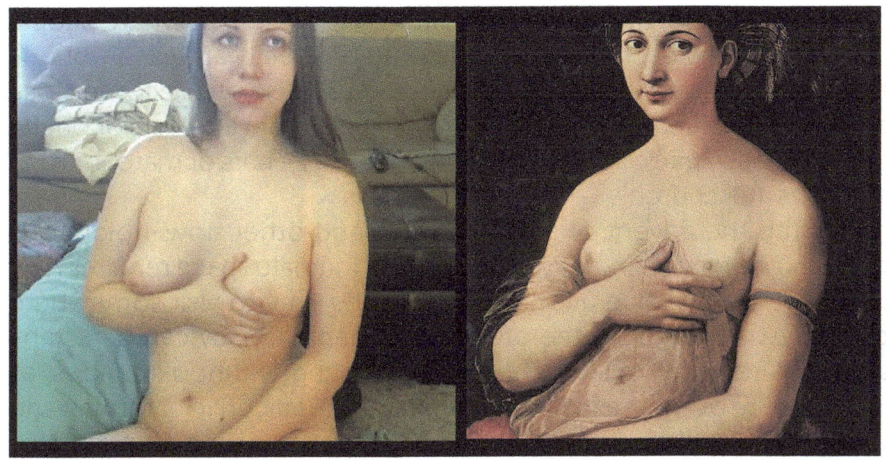

FIGURE 7.1: Addie Wagenknecht and Pablo
Garcia, *Webcam Venus* (sexcutrix as *La Fornarina*,
Raphael, 1518), 2013. Video. Courtesy of the
artists and bitforms gallery, New York.

The Proxemics of Digital Intimacy

KYLE MACHULIS

Many rituals of establishing or engaging in intimacy are tied to spaces. Whether it is hooking up at a bar, a date at a restaurant or the cinema or even a night out at a strip club, cultures create spaces where people can get to know each other and become physically and/or emotionally close (or in some spaces, to pay another person to become close quickly). These spaces may specialize in the certain interests or orientations of a specific community, such as LGBTQ+ bars or matchmaking meetups at a church.

With the advent of smart phones and other now-ubiquitous communication technology, discussion of intimate space requires new language and considerations. Why take the time to visit somewhere physical with no guarantee of reward, when intimacy can be established via remote social technologies? Dating and hookup apps, as well as more overtly sexual forms of online intimacy like camming and computer-controlled sex toys, allow users to forego location requirements, albeit with a possible reduction in experiential fidelity.[1]

This chapter analyses the movement of non-verbal communication of intimate connection and interaction from the 'real' world to the digital world. Anchoring on examples of how physical spaces and rituals facilitate establishing intimacy, it will look at how those factors translate to the digital. This leads to a discussion of how understandings of virtual space, remote presence and embodiment can inform and complicate situations where physical bodies may be thousands of miles away from each other, and how people's expectations of real versus digital interaction may map to or mislead these new experiences. At each step towards greater intimacy, the gap between real-world experience and digital simulacra widens, usually to the point where conclusions are difficult to draw from comparisons. It is from this that I hypothesize that the uptake of intimate technology benefits from known rituals and spaces for users to orient themselves to. Difficulties arise when an intimate digital experience is built without accessible relations to space and bodily presence, requiring users to build their own rituals and spaces.

A good context for discussing those difficulties is the sociological theory around how people perceive their bodies in space and the extralinguistic tactics used to communicate in those spaces. Especially Hall's theories of proxemics – the study of how people socially perceive the inhabitants and features of physical space around them – and their modern day uses, unveil the complications of always-on, always-communicating devices and applications. The idea of the 'digital proxemics' is a newer and more controversial topic that takes into account not only how bodies interact with each other and their environment physically but also how bodies react remotely through connected devices.

A particular case here might be dating apps and the ways they seek to reinvent intimate physical spaces and bodily interactions for digital communications. By enumerating the ritual actions of a physical meetup space like a bar or club and comparing those to the reduced, refined user interfaces of popular dating apps, we can see how technology developers translate the physical space into the virtual space of their applications. As variations in user preferences, identity and communication styles can affect how these systems are used, these analyses bring up research into user experiences, outlining requirements communities have when engaging in romantic matching while lacking the normal non-verbal communication tactics like gestures, proximity and haptics. Following the trajectory from negotiations of personal to intimate space, it focuses on the translation of physical to digital interaction between digital cam modelling services and one of their physical predecessors, the strip club. As these spaces include financial transactions as well as social/physical transactions, they require the consideration of the perspectives of both the customer (strip club patron, cam room user) and the employees (strip club dancers, cam models). Both the parties are equally affected when bodies move from being physically up against each other to being at unknown locations with differing digital interfaces for interaction. An insight into how the design of cam model sites and applications compare with the architectural layout and assigned spaces in a strip club may detail the boundaries of physical space and clarify elisions of that space in digital design. These details may skew labour requirements and expectations of participants on both sides of the screen.

Ending in the most intimate of spaces, the chapter will discuss relations of space to teledildonics, sex toys that can be remotely controlled through a computer.[2] Teledildonics is an intimate technology without a relatable, static, physical space from which developers can crib their designs from or to which users can orient their bodies and expectations. The history of the technology is rife with issues of uptake and acceptance due to design problems and misunderstandings of use. Using a combination of marketing copy and anecdotal user reports, a vague geometry of the space created by this technology can be sketched. However, the difficulty of translating one of the most intimate of human acts to the digital realm becomes readily apparent. Readers of this analysis may end up much like the owners of these products, with more questions than answers, as manufacturers and designers have struggled to create comfortable, interesting interfaces that delight (or at least, do not repulse) users. This leads to users creating their own interfaces and techniques for performing intimacy through technology.

This analysis relates to technological products and services that may assume a western, cis-heteronormative, able-bodied cultural perspective of the creation and continuation of intimate interaction. Apps and technologies are being created by developers of different cultural, sexual body identities worldwide to serve the needs of many diverse communities. Some of the issues presented here may pertain to those creations, but spaces (be they real or virtual) are handled differently by different groups, and this writing does not claim to be globally applicable.

PROXEMICS, DIGITALITY AND THE SPACE IN-BETWEEN

Speaking of non-verbal intimate interaction requires a framework to analyse bodies in spaces and the expectations of their owners. Proxemics – a field of non-verbal communication study, founded by Edward Hall in the late 1960s – is the 'interrelated observations and theories of [a person's] use of space as a specialized elaboration of culture' (Hall 1966: 1). As one of the major components of non-verbal communication – alongside facial expressions, vocal subtext, haptics, kinesics and others – physical proximity, bodily interaction and the surrounding territory in which it happens can affect interpersonal communication and interaction. While discussion of proxemics can be quite broad, covering topics of social structures related to how whole cultures build communication around contexts of bodily positioning and interaction, the focus for this chapter is on the dyadic interaction of the establishment and consummation of an intimate (though not necessarily romantic) relationships.

To classify proxemic interactions between bodies, Hall specifies four 'distance zones', each with a 'close phase' and a 'far phase' (Hall 1966: 116–19), delineated not only by distance but by the sensory experiences involved. Those zones include:

Public Distance Zone

- Far Phase (25 feet or more): Used for large-scale public address; voices may need to be amplified; visual contact may be blurry.
- Close Phase (12–25 feet): Requires loud talking; participants can easily remove themselves from the interaction.

Social Distance Zone

- Far Phase (7–12 feet): Formal discourse, possibly without full trust. For instance, in business, this may be a manager behind desk/ subordinate in chair distance.
- Close Phase (4–7 feet): Informal/impersonal discourse, such as two coworkers using a closer communication distance.

Personal Distance Zone

- Far Phase (2.5–4 feet): 'Arm's Length' distance, close enough that other non-verbal communication systems can be engaged (olfactory, visual/body context).
- Close Phase (1.5–2.5 feet): The distance of hugging, or at least, grasping potential. The close phase personal distance zone is where the term 'personal space' originates.

Intimate Distance Zone

- Far Phase (6–18 inches): Romantic public engagement distance, including grasping of hands, whispering, etc.
- Close Phase (0–6 inches): The space of physical sexual interaction.

Hall's perspective of proxemics also encompasses the spaces in which bodies interact. While distance zones address bodies in relation to each other, 'territoriality' is used to refer to how someone (or a group) retains ownership of the space and objects around them. In a personal scope, one might put a coat on a seat to 'claim' it, while at a larger scope, a group or community may use aesthetic or linguistic negotiations to claim a space for themselves, such as creating spaces for the interaction of those with specific identities or orientations.

Space-design affects how bodies move and interact. Hall uses the term 'Fixed-Feature Space' (Hall 1966: 103) to refer to design patterns of static-use physical spaces (i.e. spaces with installed elements or for specific types of use). This can include geographic structures like the layouts of cities, to the flow of small, specifically utilized spaces like kitchens or stores. How one enters and leaves the area, how 'comfortable' that area is and other factors can all related to the design of the space. 'Semifixed-feature spaces' (Hall 1966: 108) have movable elements, such as chairs or other furniture, which can accommodate multiple types of activities or social situations. For instance, a set of tables may be fixed, but chairs may be moved between them depending on communal group sizes at the tables. Both fixed and semifixed spaces are important when discussing intimate social interfaces.

The theory of proxemics is built from references that predate the entry of computers into everyday life. But Hall was prescient enough to discuss corollaries to the advent of digital communication technology, stating 'if we can think of [a person] as surrounded by a series of expanding and contracting fields which provide information of many kinds, we shall begin to see [them] in an entirely different light' (Hall 1966: 115). These 'fields' were considerations of the physical realm of communication, a topic that saw much attention at the time of Hall's publications. Hall was in touch with Marshall McLuhan when McLuhan developed his theories of mass media communication.[3] The two talked at the time of how Hall's proxemics and McLuhan's theories of media overlapped. While neither was able to directly address the role of social computing, the frameworks they created for analysing communication are still applicable today.

In the past few years, researchers have brought proxemics into discourse with digital communication technology. Works like John McArthur's *Digital Proxemics* (2016), Jason Farman's *Mobile Interface Theory* (2012) and Adriana de Souza E Silva & Jordan Frith's *Mobile Interfaces in Public Spaces* (2012) are forming a new foundation for digital representations of space and options for physical bodies to fit into that representation. Exactly how this representation works is still a matter of discussion. As McArthur points out, 'digital proxemics is not an unknown area of study, but it is relatively undefined, and reflection on the role of digital technology in our spatial interactions is extremely limited' (McArthur 2016: 28). There is a wealth of current research covering topics outside of intimate space, such as large-scale locative technologies like GPS, augmentations of physical spaces through depth cameras and augmented reality and overviews of social relationships and perspectives built on global scale social networks. Instead of focusing on the larger picture, my interest lies in bringing in focus on how, through digital communication, people find and minimize distance between each other, towards inhabiting the most intimate of (possibly remote) spaces.

DATING APPS

My analysis begins by looking at places where new relationships begin: social settings for meeting new people. A typical place for a physical hookup are bars, which have steadily increased in popularity for first meetings of heterosexual couples since the 2000s (Rosenfeld, Thomas and Hausen 2019: 3). Bars are a combination of fixed and semi-fixed feature spaces that are often appropriated for relationship creation. According to Jerald Cloyd, in his sociological analysis of pairing at bars,

> [a]ction within a market-place bar usually involves a concerted attempt by members to generate some form of social encounter, whether it is just to 'meet some new people' to 'score' (have a sexual encounter), or to meet a potential spouse. (Cloyd 1976: 294)

People may go to a bar alone but also quite often go in a group of two or more friends. This allows them to have a constant space of social retreat in case of rejection or uncomfortable situations. This group may also allow one of its members to approach strangers and strike up conversations in a way that feels safer than when one does that on their own.

As a physical space, bars may contain certain fixed features, such as the bar itself, tables and furniture around, as well as the gatherings space, such as a dance floor, pool tables, video games, couches/booths, etc. These features provide flow and contexts, giving patrons choices about where and how they spend their time. They also allow patrons to establish viewpoints throughout the space, where they may 'scope out' the bar and see who else may be of interest to approach. Traversing these spaces allows for non-verbal communication via physical interaction; the chance of bumping into someone or approaching someone as people move through the space can be used as communication openers. Louder areas of the bar impose/risk certain proxemic violations because they push strangers to come into each other's intimate distance zones in order to hear each other.

Finally, bars usually stock alcohol, which acts as a 'social lubricant'. 'Can I get you a drink' is a line of a great performative potential when it comes to romance and can function as a way to decrease social and physical distance. However, alcohol often creates as many difficulties as it does affordances. While small amounts of alcohol intake can lower social barriers, too much can quickly end in disaster. Drinks can also be used to aid in the ritual of pairing, via buying them for a stranger, or to violate unspoken social agreements, via spiking or drugging them. If someone does not consume alcohol (which can happen for multiple reasons), they may find the bar space unapproachable or uninhabitable.

Technology has continually tried to augment and improve upon the rituals of courtship. Video recording and replication brought about video dating, where participants would visit a studio, film an introduction of themselves, then would be given access to VHS tapes containing introductions to possible dating choices. During the 1980s and 1990s, video dating allowed people to meet in non-interactive settings, but with more information being shared than via newspaper ads (Ahuvia and Adelman 1992). The rise of internet usage and the worldwide web in the 1990s brought about dating websites, removing the require- ments of participants having to show up to a studio and acquire phys- ical media to make choices and replacing it with a simple online form and profile. With the advent of the smart phone and mobile software applications (or 'apps') in the mid-to-late 2000s, dating apps have again optimized the process, providing an interface custom made for smart phone usage. Unlike prior technologies, dating apps have become the most popular way (physical or digital) for new couples to meet (Rosenfeld, Thomas and Hausen 2019).

The transfer of a hookup space to smart phones has effectuated a transfer of the entire hookup choreography. As McArthur points out, 'digital communication has the ability to alter the ways [that the prox- emic classification of] distances are perceived in space and to recreate the zones of interpersonal distances for shared spaces created online' (McArthur 2016: 38). Dating apps are an example of this type of thinking.

A user's decision to contact someone of interest physically would involve evaluating them from a distance then walking up to them. Dating app replaces this with a 'swipe' gesture: swiping right for acceptance and to continue conversation with the current profile, left to reject and move on to the next profile. This gesture was built in a way that mimics the space and presence of being at a bar. Accord- ing to Brooke Hollabaugh – senior product UX designer on Tinder, a popular dating app –

> [t]he swipe itself mimics real life. Glancing at that cute guy in a bar, you sort of swipe right or left with your eyes and make a deci- sion if you like them. If they make eye contact back, you connect. Translating that into an app, Tinder gives you that same variable reward feeling you get when you receive a match. The left swipe removes the fear of rejection you face in that same scenario at a bar, as well as the guilt you feel by rejecting someone. They never have to know you swiped left. (Faller 2018: n.pag.)

Alongside the replacement of searching and selection via app comes the replacement of the requirement to be in a certain physical place to perform these actions. Dating apps can be used anywhere that the user can have their smart phone, meaning that the rituals of dating selec- tion can continue at work, at home, or wherever else a user might have time to engage with the app. The new dating app place no longer has the limitations of physical space that can only hold a certain number of bodies, so exhausting the amount of choices available is no longer an issue. While someone at a bar may feel that a night out is a failure due to the lack of anyone else in attendance that they are interested in, a dating app allows that person to scroll seemingly infinitely through as many profiles as they please.

> I heard a comedian once say that it's like we're basically sitting here with a bar in our pocket – we're sitting in a bar full of single people, looking on our phones for other single people that aren't there with us', Pete Vasconcellos, bar director at New York's The Penrose, says. 'It's really weird and backwards, but that's the new normal.' (Latterner 2018: n.pag.)

Compared to in-person meeting, dating apps reduce the range of prox-emic interactions available. The traversal of multiple proxemic zones simplifies to a version of the Public Distance Zone that is unique to a dating (or social media) apps. Profiles provide a tailored view of a person but severely limit much of the non-verbal communication that might otherwise happen in a physical environment. This fundamentally changes how users interested in each other may access the other's suitability and attractiveness (Fiore et al. 2008). As meetings are arranged on the app, the first physical encounter will normally be planned in a space that partici-pants are comfortable with. Negotiating this meeting necessitates prox-emic comfort within the social or personal distance zones.

   This reduction of choices can be positive for some people. As Cloyd points out, '[m]uch barroom behavior seems disjointed and confused, with loud music in the background, possibly some members dancing, and many individuals making numerous and often fleeting contacts with others' (Cloyd 1976: 294). This situation may not be ideal for those who are introverted, prone to overstimulation or are disa-bled and may not be able to easily make their way around a bar easily. (Though, according to some researchers [Ladau 2017; Saltes 2013], online dating should not be seen as a panacea for the disabled.) The filtering of interaction and communication types in dating apps may benefit some of these groups.

   Even with the move to dating apps, bars still thrive, as they now become meeting places for people who have connected via apps, instead of places where the connections are created. However, apps have changed the proxemics of the bar space:

> 'It's a day that you know you can get a seat at the bar because it's not going to be that crowded, but our bar is literally 100 percent Tinder dates on Tuesday', Vasconcellos says. 'They all show up

at 9 o'clock and they all have one drink and sit there and nurse it all evening... Those early nights of the week are very Tinder-heavy.' (Latterner 2018: n.pag.)

CAM MODELLING

While some arrangements of proxemics for situations of intimacy seek the goal of a social (and possibly romantic) relationship, others take the form of a combination of business transactions and sexual, emotional and/or therapeutic engagement. In the physical world, strip clubs provide a way for customers to subvert proxemic norms via monetary exchange. Cam modelling (or camming) exists as a digital equivalent, letting models engage anonymously from unidentified locations via digital cameras streaming to the internet. Customers can be similarly anonymous, viewing the live video streams of models while transferring them money over the internet to have models perform certain acts, from disrobing to more graphic sexual interaction. Camming has existed as a business for over two decades and is continually remaking itself based on the newest and latest technological advances. To quote a *New York Times* description of a cam model's job:

> Lacey is a cam model. She performs one-woman sex shows, often from her house, though she has performed in a car, on a hiking trail, and once at an airport. The action is captured by a small, inexpensive camera clipped to the top of her laptop, and made available to anyone who visits a Web site called MyFree-Cams. (Richtel 2013: n.pag.)

This description shows the deep proxemic flexibility of using a modern digital media distribution platform as a sexual interface. The body is not required to exist in a space carved out for sexuality, but rather, if the available internet connection is fast and video can be sent, any space can be adapted for sexual needs. This flexibility does not come without costs. Introducing always-on, always-connected technology to the deeply affective realm of sexuality can blur the borders of communication, consent and the body. Comparing the real spaces of strip clubs to the virtual spaces of cam modelling can identify where and how these complications occur. Via its user interfaces, cam modelling in some ways mirrors the use of space in strip clubs, either through necessity or skeuomorphisms, while the additions of cam model customizations of space and accessibility add dimensions that would be difficult to match in a public, shared physical space.

Considering the evolution of spaces for intimacy-as-trade provides a lens to evaluate their current state. Spaces where people take their clothes off for money have existed for centuries, albeit having a specific place to do so is a more recent development. Staying within the confines of the past couple of centuries, strip clubs developed from the burlesque ideal of a 'saucy blend of music, comedy, clever social satire, and bold sexuality piquing the interest of audiences' (Fargo 2019: n.pag.), up through the pole dancing of the 1960s (Hall 2018) and lap dances of the 1970 and 1980s (Steinberg 2004), putting the body of the worker (the dancer) nearer the bodies of the clients (the audience) as development continued. The proxemics of strip clubs changed over time, as societal norms and laws allowed.

When analysing space and interaction in exotic dance clubs, we can rely on the normal models of proxemics. Patrons pay for the reduction of distance into intimate space. The more money that is spent, the less space between the worker and the customer (with limitations based on the comfort the worker, legal restrictions, etc.), possibly with a longer duration for the interaction. Violations of these agreed-upon norms are handled via physical violence enacted by bouncers at the club.

The embodied experience of the patron is contextualized on their reasons for visiting the club. Where the patron is located in the club can affect the expectations of workers, as 'the closer a patron is to the main stage, the more frequently he is expected to tip; those on the tip row are expected to tip every dancer on stage' (Erickson and Tweksbury 2000: 278). How patrons situate themselves in the club may speak to their goals in attending. According to Kim Price-Glynn (2010: n.pag.):

> While some men are entering the club to see and experience their sexuality in ways that may be unavailable in their intimate lives; others seek to experience sexuality collectively with an emphasis more or less explicitly on male bonding; still others seek to experience sexuality through a sense of dominance as men and heterosexuals.

As shown in ethnographic research of clubs, there are multiple areas that can be occupied in a club that change the context of interaction between the customers and the workers. Expectations and actions can change depending on whether the customer is near the stage, sticks near the walls, etc. Much like bars from the dating example, clubs are laid out with many completely or mostly fixed feature areas. The stage where the dancers work, tables throughout the club and the club bar all function as social areas like a regular bar, though with different contexts, as patrons expect a different type of interaction in the space. As workers are expected to interact with patrons in an 'in-character' manner while in areas where business may be transacted, they have areas where they can 'break character' as they take breaks.

> The space in which our interview took place was significant because in this case, she was physically outside the club. She talked with me as if she were meeting me at a party instead of as of a potential client – though at times she reverted back into her sexualized spectacular character from the stage, perhaps with the expectation that we would become potential clients. Her life as a stripper seemed wrought with this normal versus sexualized dichotomy, as if she did not know where to put herself. (DeMarco et al. 2010: 5–6)

As a digital equivalent to a strip club, cam modelling (or camming) is a younger industry, establishing itself in the late 1990s. Starting as live images updated at three to five times per second, camming capabilities have grown alongside streaming technology, with modern-day plat-forms using real-time video feeds at high resolutions, at times broad-casting to thousands of viewers.

The transaction system for camming resembles that of a live adult entertainment venue, like a strip club or peep show, as accessed through a digital portal. Websites such as chaturbate.com or myfreecams.com provide a central hub for models to post their streams and attract custom-ers, offering a monetary transaction system, usually in the form of virtual 'tokens'. Models can set certain token rates for which they will perform certain actions. For instance, removing a piece of clothing may cost the equivalent to a few dollars, while a graphic sex act may be tens or even hundreds of dollars, depending on the act, the model's popularity and other factors. The goals of the platform can change depending on the model, as Eduardo Martins points out, 'There are users who broadcast randomly and do not see the platform as a source of income, while others face it as an actual occupation and put a lot of effort to deliver quality erotic content and earn from it' (Martins 2019: 2).

The interface of camming sites creates what could be referred to as 'asymmetric proxemics'. For the model, all that can be seen is a text chat window with the responses and conversation of viewers. This chat window may have certain interface elements that help the model identify users, such as colour coding to which users have tokens to give (some sites show the names of users with no tokens as a grey colour, leading to models referring to them as 'greys'). For viewers, they have full video and audio streams available through which to view the model. To use Shannon's theories of communication and informational bandwidth

(Shannon 1948), the channel capacity of the viewers watching the model is far greater than the capacity of the model's perspective of the viewers. Compared with the strip club, where dancers and customers can see and interact with each other through multiple types of non-verbal communication, a cam model must use both verbal (speaking that is picked up in the audio stream) and non-verbal (bodily movement) communication to entice them to spend more tokens, while the only input returned to the model is text chat and token amounts.

The interface of a cam room presents a unique type of fixed feature space. While models can customize the space shown on the video stream, along with what they wear (or don't), they must stay within the boundaries of the camera frame. In clubs, dancers have choices in clothing, but not environment. However, the environment is normally large enough to walk around and may involve multiple stages or areas within which to interact with patrons. Cam rooms require movement of the camera to change the scenery, which may be difficult or impossible depending on the current situation and actions of the model. Similarly, patrons of a cam room are kept to whatever chat interface is provided by the cam website. While they may be able to customize their names and possibly chat colours (for patrons that have paid for accounts), other expressions of individuality must be made through the chat interface. The ability to choose diverse and accommodating working environments makes camming (both modelling and viewing) popular for those who may face accessibility issues in actual clubs. Disabled cam models have reported that camming gives them an income and freedom they would not normally have access to (Brasseur and Finez 2020: 222).

As the digital portrayal of space and bodies in a cam room does not allow for the model to personally engage with individual patrons, there is also no way for the model to take a break when on cam. As dating apps present users with unlimited profiles to interact with, a cam site interface allows virtually infinite choices of cams to watch. A cam model taking a break from their work usually means shutting off the video stream. Unlike a strip club, where dancers will rotate working throughout the evening but are usually on schedules, every other model working on a cam sites can be viewed as competition. Any time away from the camera, or time when a stream is shut down, means that a model must reestablish their viewership starting from near zero. Repeat customers may join and being to fill out a room once a model announces they are on stream, but attainment of new customers requires working for far longer shifts than a club situation would require.

With camming, embodied engagement for the viewer changes as much as it does for the model. Engagement for the strip club patron, as explained earlier, provides a straightforward embodied experience, as they are physically at the club, interacting with workers. They can move about the space without needing an intermediate communication medium but must work within the rules of the venue or face ejection. Cam viewers must use a computer to interact with models, meaning that they have more freedom to act physically but less access for communication. While a patron at a club can use non-verbal communication to show interest in a particular dancer, cam viewers must navigate and interact with a technical interface to reach their goal, similar to how Patrick Keilty describes the embodied engagement of browsing for pornography:

> the interface of online pornography necessitates delays: logging on, finding a site, scrolling up and down/left and right, opening and closing windows, clicking forward/ backward, and pushing the refresh button. (2016: 67)

While this interface is less engaging than being immersed in a club environment, it provides certain allowances for the viewer. They may wear as much or as little as they please and manipulate themselves in ways that would otherwise be forbidden in the public club environment.

The change in engagement for the cam viewer extends beyond their passive intake of the situation and into the interactive aspects of personal transactions. Tipping a dancer in a strip club is a way to reduce space between the dancer and the person who is tipping, as the dancer has to come over to retrieve the tip and, depending on the amount, may take requests, spend more time near the customer or provide other rewards based on the transaction. While other patrons of the club may watch the interaction, it will be from a noticeable distance from the tipper. As described by Erickson and Tewksbury, 'the [strip club] customer may dictate the nature, and often the course, of the interactions because the dancer is both obligated and financially motivated to cooperate with the direction of the customer in defining the interactions' (Erickson and Tweksbury 2000: 273). With cam modelling, unless the model goes to a private online venue with a single member, most tips are rewarded in a public setting. One person tipping a cam model means all users watching, both tipping and non-tipping, are rewarded. The tipper may request certain acts that they would enjoy, or the model may refer to them in person, but within the camming space, there is no way for the tipping patron to be rewarded in real time in a way that does not involve the rest of the viewers.

The world of digital exhibitionism via camming creates a complicated social model for both performers and viewers, as McArthur points out:

In the digital world, a sociopetal experience that is also proxemofugal[4] is perhaps the experience of the digital exhibitionist. The exhibitionist is oriented toward the camera, toward the audience, but is projecting her present environment out through broadcast. Even though the exhibitionist is physical present in space, she is not connecting to physical space. Instead, she is broadcasting her space to be experienced by and connected to others. In contrast, a sociofugal experience that is perhaps also proxemopetal in the digital world is the experience of the digital voyeur. The watcher is not oriented toward social interaction, but only toward entertainment. The voyeur connects with the space created onscreen, despite the fact that such space is not his to occupy. While the voyeur may connect with an on-screen other, such connection only occurs in his mind. (McArthur 2016: 115–16)

TELEDILDONICS

Creation of digital equivalencies to intimate physical spaces is only a subset of what is possible with intimate technology. While people can meet, converse and view each other online, we can now also relay touch, both informative and intimate, through remote means. Edward Hall spoke of the human want to extend beyond bodies, noting that 'man has developed extensions for practically everything he used to do with his body' (Hall 1959: 79). By combining haptic (touch) technology with digital communication strategies like audio/video streaming and text chat, digital extensions of the body can reach into intimate space.

Tracing the formation of the term 'teledildonics' back to its roots leads to 1974, when Ted Nelson coined 'dildonics' as part of an article on an inventor working with acoustically actuated haptics (Nelson 1974). At the time, Nelson used the term to refer to sexualization of a multi-use device. Howard Rheingold built on this term in 1990, coining 'teledildonics' to refer to remotely actuated intimacy through technologically controlled hardware (Rheingold 1990).

According to Rheingold, '[t]eledildonics is inevitable given the rate of progress in the enabling technologies of shape-memory alloys, fiber-optics, and supercomputing' (Rheingold 1990: 53). The progression from ideation to actuation turned out to be more mundane. Products like the Safe Sex Plus started to appear in the late 1990s, with the first boom happened in the mid-2000s as companies like Highjoy and Sinulate came to market alongside the proliferation of in-home broadband in the United States. This hardware allowed users to remotely control vibration speeds in different form factors, from dildos to sleeves.

It also required a wired connection to a desktop computer and control with a mouse and/or keyboard, making usage an awkward and sometimes challenging situation. The advent of the smart phone in the late 2000s, combined with wireless communication technology such as Bluetooth, allowed digitally connected intimate devices to remove their cable tether while giving users a more natural interface by using phone touch screens instead of mice and keyboards. This led to another boom in the industry around 2013, with companies like Lovense and Kiiroo bringing products to market and finding traction among a new set of consumers who may not have had the technological know-how to use previous iterations of these devices.

Computer-controlled sex toys can be used in solo situations, through video games, movies and other media, but many users seek out the technology to establish and fulfil the dream of intimate dyadic interaction via remote technology. Manufacturers use this goal in their advertising, with ad copy touting the benefits of remotely connected intimacy:

> Connectivity isn't a substitute for having your partner beside you, but for those who have to endure weeks or months apart, it's the best/easiest way to close the intimacy gap (when watching Skype masturbation isn't enough anymore). (Lovense, 2017b: n.pag.)

The social and emotional connection between participants plays an extremely important part in remote intimate interaction. Teledildonics augments that connection by adding the physical aspect to touch. The state of current technology means that the feeling of touch provided by teledildonic hardware will not be the same as skin-to-skin contact, but it still provides a touch experience triggered by another person.

Hall points out that '[a person's] feeling about being properly oriented in space runs deep' (Hall 1966: 105), and in this quote lies a prevalent issue with teledildonics technology as it stands now. Teledildonics seeks to augment or replace an act, one that may otherwise be impossible due to geographic separation of the participants. It does so with no reference to the space in which that act should happen, leaving participants to figure out how to orient themselves with the technology and each other. An act that is normally intimate, happening on a bed or other furniture, within the Intimate Close Phase zone of proxemic definitions, now must be communicated and actuated through whatever interfaces are available on a computer or smart phone. Touch turns to movement of a slider bar, motion of bodies is translated to the quantized computational playback of command patterns, communication can be limited or obscured based on technology and network connection quality and space must be approximated via whatever means are available. Video and audio chat replace body-to-body closeness, and participants must perform within the confines of a webcam as well as the restrictions of how the teledildonic hardware works with their bodies. Distraction from the task at hand can be caused by the hardware itself. Motor sounds and desynchronization of movement between the toy and what may be happening on video or audio can invoke a feeling of dissociation from the intimate events, and the haptic experience the hardware provides cannot currently mirror anything resembling the actual physical experience of sexual intercourse.

In response to these complications, toy manufacturers host usage explanations and advice on their websites. For instance, Lovense specifically addresses couples who feel that 'watching Skype masturbation isn't enough anymore' (Lovense 2017a, 2017b). This line gives us a perspective into the potential users of this technology: they may have already engaged in remote sexual activity through purely auditory and visual contexts, may already be familiar with the requirements for connecting to each other via digital communication services and are looking to add something new to this interaction. In this case, the issues of bodily accommodation and orientation may be something they have already dealt with. Like cam models, the participants have learned to exist within the webcam frame when interacting with each other. The added awkwardness of teledildonic hardware usage and control may be seen as just another obstacle to be cleared on the way to greater remote intimacy.

Manufacturers of these products are aware of these issues with usage and space, and their thoughts on the subject are reflected in the copy on their websites and advertising. For instance, Lovense states that '"wearable" sex toys (aka teledildonics) are vibrators that will stay inside of you while you go about your day' (Lovense 2017a, 2017b), changing the territoriality of intimacy from something that happens in a private space to something that can happen anywhere. The company touts the qualities that allow their products to recontextualize any space, public or private, into an intimate, sexual space. Images on the website depicting public use of toys are captioned with 'Discreet Public Play', and product pages contain copy like 'Nearly Silent When Inserted; Use without worrying about roommates, family, or strangers in public hearing it!'. Not only does this ad copy sell the product as a way to increase physical sexual pleasure, but it also recognizes and lauds its ability to carve intimate, dyadically proxemopetal spaces out of otherwise shared public territories.

While the technology continues to find footing in appealing to those in long distance relationships, cam models have found use for teledildonics as part of their labour. Starting in the early 2010s, models began to use toys that would vibrate whenever certain sounds were played. These sounds were triggered by patrons giving tokens to the model. As technology progressed, these toys moved from using sound (which was difficult to isolate and control) to being completely computer controlled. This addition allowed models to give patrons something that strip clubs could not: direct sexual interaction from tips. Tip amounts change the speed the toy vibrates at or the length of time for which it vibrates. While the bodies of the patrons and performers were separated, this re-establishes a direct physical interaction from a patron's action. The complexities of this arrangement are explained by Martins:

> The number of individuals in charge of the functioning of [the sex toy] thus goes from a single person to thousands at the same time, making the performer completely susceptible to tippers' commands. The vibrator itself is also involved in the management of the situation, for to feel pleasure, cammers and viewers must subject themselves to its mode of operation, and consent to its functioning configurations and set of adjustments. Thus, the sexual activity is modulated by all participants – cammer, tipper, vibrator – and [the vibrator] plays the people who use it as much as it is played by them. (Martins 2019: 6)

While the technology lacks a relational physical space, the users of teledildonics still create proxemicly interesting relationships with each other through intimate haptic communication. By fulfilling requirements of intimate connection in a remote relationship, augmenting labour or subverting social norms for excitement, teledildonic technology allows the redefinitions and recontextualizations of digitally created spaces. It does this instead of and in contrast to synthesizing familiar physical spaces.

CONCLUSION: SPACES WITHOUT BODIES

The interaction of physical bodies plays a significant role in the establishment and engagement of intimacy. They are the vessels through which emotional communication comes to being. Digital intimacy diminishes or even removes the requirements of the physical body from these situations, augmenting or replacing it with communication capabilities. Whether these capabilities are enough to create a fulfilling experience of digital intimate interaction depends on the users and the contexts of their situation. Nast and Pile explain how proxemics can relate these spaces and bodies, even though neither may be physical:

> Proxemics teases out the simultaneously fixed and fluid nature of spatial arrangements by articulating the sense that networks shift, alter and stabilize around effects of power, meaning, subjectivity and objectivity. These effects are both bodily and spatial, but are rarely explicit or open or conscious. Instead, they are felt; thought through the body rather than the mind. Thus, proxemics describes an 'unconscious' relationship between the body and spatialities. (Nast and Pile 1998: 304)

What has been covered in this chapter serves as a starting point for analysing how strategies for online intimacy synthesize, simulate and create spaces where proxemics may not resemble the real world. Dating apps, camming and teledildonics make up a subset of ways people become close online. Comparing them to the real spaces gives insight into design strategies and user needs. As Nast and Pile put it, '[like Moebius's famous strip, proxemics moves through a space where location is indeterminate, relational, and multi-dimensional' (Nast and Pile 1998: 304), and this holds true when evaluating proxemics in the digital realm. Privacy and consent become complicated (Lewis 2017), translation of non-verbal communication and identity may require new or different skills (Waskul 2003) and models of labour can evolve in new and unexpected ways (Martins 2019). While physical bodies and spaces may sometimes be absent from digital experiences, the ideas of bodies and spaces remain integral to making digital experiences understandable and intimate.

References

Ahuvia, Aaron C. and Adelman, Mara B. (1992), 'Formal intermediaries in the marriage market: a typology and review', *Journal of Marriage and Family*, 54:2, pp. 452–63.

Beckhaus, Steffi and Lindeman, Robert W. (2011), 'Experiential fidelity: Leveraging the mind to improve the vr experience', in S. Coquillart, G. Welch and G. Brunnett (eds), *Virtual Realities*, Vienna: Springer, pp. 39–49.

Brasseur, Pierre and Finez, Jean (2020), 'Performing amateurism: A study of camgirls' work', in S. Naulin and A. Jourdain (eds), *The Social Meaning of Extra Money*, New York: Palgrave MacMillan, pp. 211–37.

Cloyd, Jerald (1976), 'The market-place bar: The interrelation between sex, situation, and strategies in the pairing ritual of homo ludens', *Urban Life*, 5:3, pp. 293–312.

DeMarco, Tony, Kramer, Martha, Lanz, Ali, Looze, Monica, Sadowski, Cassy and Zhu, Jennie (2010), 'The strip club as a simulacrum: An analysis of illusion, perception, and power', *The Strip Club as a Simulacrum*, http://www.monicalooze.com/an-ethnography-of-a-strip-club. Accessed 18 November 2019.

de Souza e Silva, Adriana and Frith, Jordan (2012), *Mobile Interfaces in Public Spaces*, New York: Routledge.

Erickson, David John and Tweksbury, Richard (2000), 'The gentlemen in the club: A typology of strip club patrons', *Deviant Behavior*, 21:3, pp. 271–93.

Faller, Patrick (2018), 'For the love of ux: Tinder's product designers talk user-centered design for emotional experiences', Adobe.com., 14 February, https://xd.adobe.com/ideas/perspectives/interviews/love-ux-tinders-product-designers-talk-user-centered/. Accessed 23 January 2020.

Fargo, Emily Layne (2019), 'Beginnings of burlesque', *Loose Women in Tights*, 19 September, https://library.osu.edu/site/burlesque/. Accessed 12 January 2020.

Farman, Jason (2012), *Mobile Interface Theory: Embodied Space and Locative Media*, New York: Routledge.

Fiore, Andrew T., Lindsay Shaw, Taylor, Mendelsohn, G.A. and Hearst, Marti (2008), 'Assessing attractiveness in online dating profiles', in Anon. (eds), *Proceedings of the SIGCHI Conference on Human Factors in Computing Systems (CHI '08)*, New York: Association for Computing Machinery, pp. 797–806.

Hall, Destynnie (2018), 'Origins and history of pole dancing', *Polepedia*, 1 July, https://polepedia.com/origin-history-pole-dancing/. Accessed 20 January 2020.

Hall, Edward T. (1959), *The Silent Language*, Garden City, New York: Doubleday.

Hall, Edward T. (1966), *The Hidden Dimension*, New York: Anchor Books.

Jones, Angela (2016), '"I get paid to have orgasms": Adult webcam models' negotiation of pleasure and danger', *Signs: Journal of Women in Culture and Society*, 42:1, pp. 227–56.

Keilty, Patrick (2016), 'Embodied engagements with online pornography', *The Information Society*, 32:1, pp. 64–73.

Ladau, Emily (2017), 'Playing the online dating game, in a wheelchair', *New York Times*, 27 September, https://www.nytimes.com/2017/09/27/opinion/online-dating-disability.html. Accessed 7 March 2020.

Latterner, Tim (2018), 'Dating apps have taken all the romance out of going to bars', *VinePair*, 26 April, https://vinepair.com/articles/its-the-new-normal-digital-dating-apps-are-transforming-the-bar-industry-irl/. Accessed 8 February 2020.

Lewis, Sarah Jamie (2017), *Queer Privacy: Essays from the Margins of Society*, International: lulu.com.

Lovense (2017a), '12 orgasmic sex toy games – have fun with your teledildonics', *Lovense*, 19 November, http://www.lovense.com/sex-tips/sex-toy-games. Accessed 27 February 2020.

Lovense (2017b), 'Teledildonics 101 – learn about emerging sex tech & orgasm gizmos', *Lovense*, 28 October, https://www.lovense.com/teledildonics. Accessed 6 March 2020.

Martins, Eduardo E. B. (2019), 'I'm the operator with my pocket vibrator: Collective intimate relations on chaturbate', *Social Media + Society*, 5:4, pp. 1–10.

McArthur, John A. (2016), *Digital Proxemics: How Technology Shapes the Ways We Move*, New York: Peter Lang.

Nast, Heidi and Pile, Steve (1998), 'Every-DayBodiePlaces', in H. Nast and S. Pile (eds), *Places through the Body*, London: Routledge, pp. 405–16.

Nelson, Ted (1974), *Computer Lib/Dream Machines*, Self-Published.

Price-Glynn, Kim (2010), *Strip Club: Gender, Poweand Sex Work*, New York: NY University Press.

Rheingold, Howard (1990), 'Teledildonics: Reach out and touch someone', *Mondo 2000*, Summer, pp. 52–54.

Richtel, Matt (2013), 'Intimacy on the web, with a crowd', *New York Times*, 22 September, https://www.nytimes.com/2013/09/22/technology/intimacy-on-the-web-with-a-crowd.html. Accessed 13 January 2020.

Rogers, Everett (2000), 'The extensions of men: The correspondence of Marshall McLuhan and Edward T. Hall', *Mass Communication and Society*, 3:1, pp. 117–35.

Rosenfeld, Michael J., Thomas, Reuben J. and Hausen, Sonia (2019), 'Disintermediating your friends: How online dating in the United States displaces other ways of meeting', *Proceedings of the National Academy of Sciences of the United States of America*, 116:36, pp. 17753–8.

Saltes, Natasha (2013), 'Disability, identity, and disclosure in the online dating environment', *Disability & Society*, 28:1, pp. 96–109.

Shannon, Claude E. (1948), 'A mathematical theory of communication', *The Bell Systems Technical Journal*, 27, pp. 379–423, 623–56.

Steinberg, David (2004), 'Lap victory: how a DA's decision to drop prostitution charges against lap dancers will change the sexual culture of S.F. -- and, perhaps, the country', *SF Weekly*, 8 September, http://www.sfweekly.com/2004-09-08/news/lap-victory/. Accessed 28 January 2020.

Waskul, Dennis (2003), *Self-Games and Body-Play*, New York: Peter Lang.

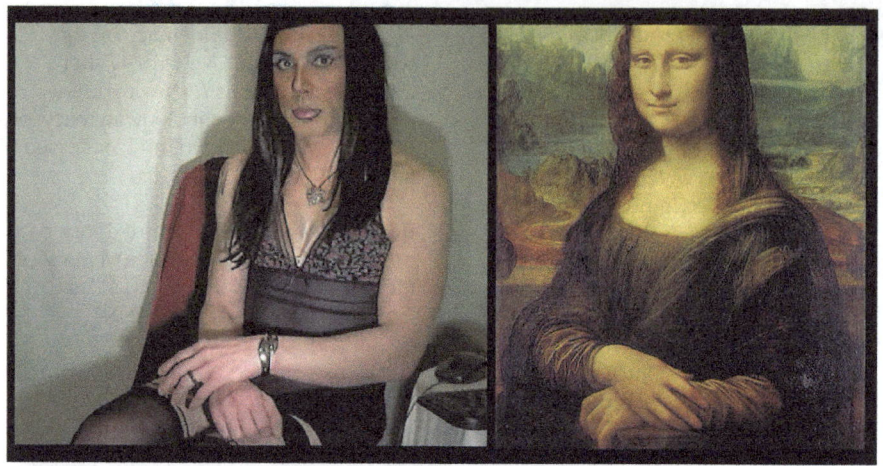

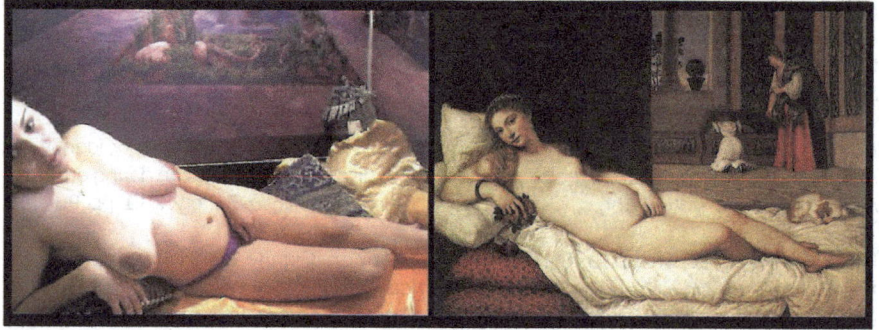

FIGURE 7.2: Addie Wagenknecht and Pablo
Garcia, *Webcam Venus* (ricadoll as *Mona Lisa
–La Gioconda*, Leonardo da Vinci, 1503), 2013.
Video. Courtesy of the artists and bitforms
gallery, New York.

FIGURE 7.3: Addie Wagenknecht and Pablo
Garcia, *Webcam Venus* (kimisquirtx as *The Venus
of Urbino*, Titian, 1538), 2013. Video, 02:41.
Courtesy of the artists and bitforms gallery,
New York.

FIGURE 7.4 (page 157): Addie Wagenknecht
and Pablo Garcia, *Webcam Venus* (lollyroo as
Milkmaid, Vermeer, 1658), 2013. Video. Courtesy
of the artists and bitforms gallery, New York

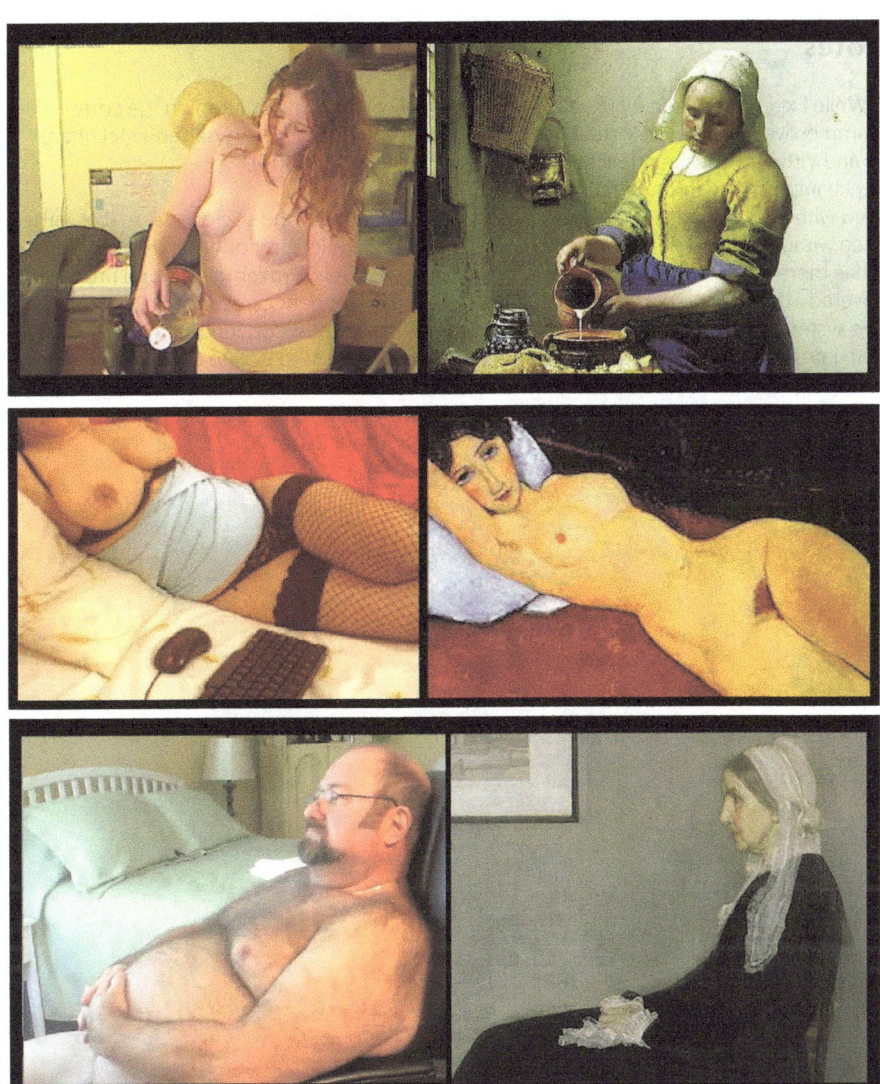

FIGURE 7.5: Addie Wagenknecht and Pablo Garcia, *Webcam Venus* (boobz_4_play as *Reclining Nude*, Amedeo Modigliani, 1917), 2013. Video. Courtesy of the artists and bitforms gallery, New York.

FIGURE 7.6: Addie Wagenknecht and Pablo Garcia, *Webcam Venus* (frogmann as *Whistlers Mother*, James Abbott McNeill Whistler, 1871), 2013. Video. Courtesy of the artists and bitforms gallery, New York.

Notes

1. While Experiential Fidelity usually refers to immersive systems like Virtual Reality and Virtual Worlds in publications like Beckhaus and Lindeman (2011), here it is used with apps that may replicate experience without full audio/visual immersion. While there are many fruitful topics to be covered in intimate experiences via immersive systems, this chapter will centre on non-immersive interfaces, such as social dating apps and sexual interfacing services.

2. As coined in (Rheingold 1990), via Dildonics from (Nelson 1974).

3. For more information on the correspondence between Hall and McLuhan, see Rogers (2000).

4. Sociofugal and sociopedal spaces either keep people apart or bring them together, respectively. Proxemofugal and proxemopedal experiences, as McArthur uses the terms, refer to concentrating on the digital environment to isolate from the physical, versus using digital technology and environments to augment physical spaces. (McArthur 2016: 70–71).

8
Touchless Embraces

Part of the sensitive media paradigm – according to which media devices feel with us and we feel with the devices – is our sensory and motor entanglement with the media in which the media determine our sensual and physical responses. Perhaps, the most interesting in this context is the transformation of touch brought up by the emergence of haptic technologies. When we look at the prevalence of tactile input hardware (e.g. touchpads, touchscreens) and translate it into our daily media interactions through touch, we may say we are dealing with the new tactile awareness that develops alongside the increase of touch-operated interfaces. At the same time, we can talk about the crisis of touch caused by the distancing effects of virtuality, enhanced by the onset of tactile and vibrotactile devices by means of which we transfer sensation to another human or device.

Tom Galle and Moises Sanabria's *VR Hug* pictures this crisis with an almost iconic image of a couple in a tight embrace, each of them wearing a VR headset. An allusion to mediated hookup so many of us are choosing these days, *VR Hug* poignantly conveys the shift in sensory modality and the replacement of touch with mediated intangibility. The work brings out some evident ambivalence of virtualization – its distancing function and stimulating potential. As the VR headset anchors the couple outside of the physical contact, they disorient themselves from each other to find pleasure in a non-tactile experience. There is something about those augmented media worlds that lure us to abandon our offline togetherness. There is also something about us – a curiosity, insatiability or convenience – that makes us pursue whatever there is outside of the 'face to face'. Perhaps, the old reality of bodies is no longer enough. Indeed, the media devices offer sensations and experiences that the physical reality within our reach cannot provide. And we all carry the awareness of their dislocating pleasures, knowing that with the media we are always 'here and somewhere else' – both at the same time. These bilocation and bi-sensation are part of our digital condition.

But if love is blind, do those augmented visions make it less obscure? Does mediated contact flesh out sensations better than a traditional body-to-body encounter? Like in Magritte's 'The Lovers' (that Galle and Sanabria's work bears a semblance of), the couple in *VR Hug* is affirming the elusiveness of a physical union. They are also affirming the imaginative aspects of intimacy and intimate interactions. Virtual reality is no less real than the so-called normal reality. Analogically, virtual sex or virtual love are essentially no different from love and sexual normality we are familiar with. If much of what we understand as love is an effect of some social code of communication and emotional formation, virtuality may simply be a different dimension of the already chimerical experience.

VR Hug

TOM GALLE AND MOISES SANABRIA

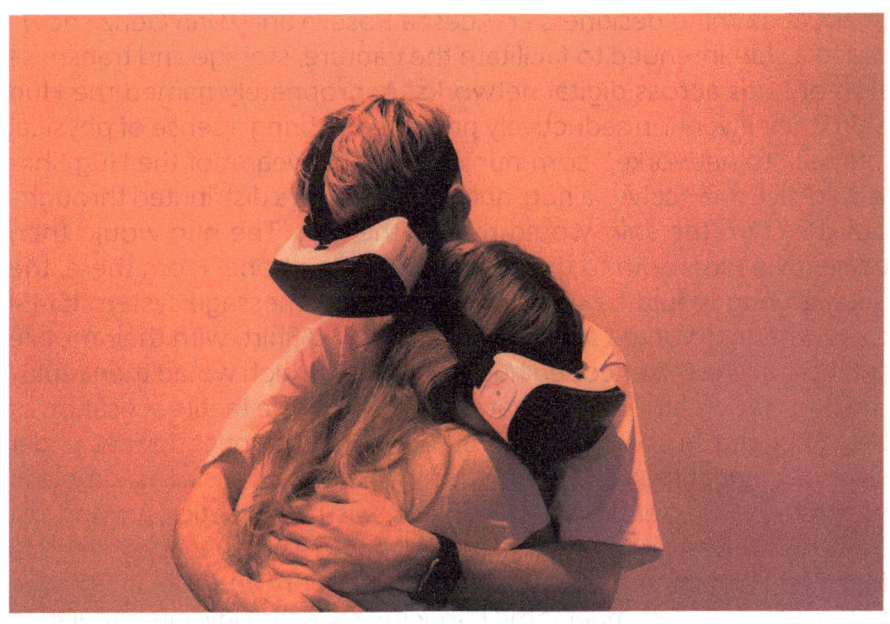

FIGURE 8.1: Tom Galle and Moises
Sanabria, *VR Hug*, 2016. Acrylic print.
Courtesy of the artists.

Virtual Hugs and the Crises of Touch

DAVID PARISI

In 2002, clothing designers Francesca Rosella and Ryan Genz show-cased a shirt intended to facilitate the capture, storage and transmission of hugs across digital networks. Appropriately named the Hug Shirt, their invention seductively promised to bring a sense of physical intimacy to networked communication. The wearer of the Hug Shirt would give themselves a hug, and, using sensors distributed through-out the shirt, the shirt would record the hug. The hug would then be sent via Bluetooth to the wearer's mobile phone. From there, the datafied hug would be transmitted via short message system (SMS) to another networked subject wearing a Hug Shirt, with their mobile phone relaying the encoded hug to the shirt, which would then replay the hug via actuators that vibrated to simulate the tactile sensation of being hugged. In comparison to other wearable haptics devices under development at the time, the device was modest in its aims, simple in its function and intentionally elegant in its visual aesthetics. Arriving just as the never-realized dreams of 1990s virtual reality were beginning to fade, the Hug Shirt provided a refreshingly bright and light alternative to the dark, bulky, cumbersome haptics vests and bodysuits circulating in the popular imagination during the previous decade. Shortly after its debut, the Hug Shirt caught the attention of the popular press. *Time* named it one of their inventions of the year in 2006, hailing the Hug Shirt as a way to revert the meaning of public display of affection (PDA) from the then-in vogue Personal Data Assistant back to the more inti-mate PDA. A year later, it was featured at *WIRED*'s NextFest and then in the technofetishist *World Changing* compendium soon after that. Although the Hug Shirt has only just now been released for commercial sale (available for £350), over the eighteen years since its invention, it has remained an object of ongoing fascination, showcased in a 2014 CNN segment where Rosella noted that 'people have been waiting a long time for the Hug Shirt' (CNN 2014). A 2017 *Verge* article on digi-tal hug devices occasioned by Valentine's Day teased readers with the promise that they'd soon be able to use the shirt to hug their loved ones from a distance, with no mention of the Hug Shirt's long life or present status. The Hug Shirt's historicity – the recurring rediscovery of and anticipation for the curious device that sends a hug-that-is-not-quite-a-hug – indicates a continual cultural desire for new forms of mediated touch. Simultaneously, its deferred arrival as a commercial product illustrates the challenges and limits of attempts to successfully bring technologies of networked tactility to market.

The Hug Shirt presents a fascinating case study for thinking through questions of networked affect, promising a touchless embrace that can lend materiality to expression 'sending you hugs'. Accordingly, this chapter examines the nexus of relationships between (1) the material and infrastructural processes the Hug Shirt employs to transmit hugs, (2) the semiotics of sending and receiving – of encoding and decoding – these touchless embraces, (3) the Hug Shirt's discursive construction as an object that fills a particular need in our culture of digitally networked affective communication and (4) the Hug Shirt's self-conscious arrival amidst a purported crisis of touch, a crisis attributed in part to our increasing reliance on digital technologies for interpersonal communication. The Hug Shirt provides a lens through which we can view the history of haptics technologies more generally, showing how these devices have been articulated as a means of ameliorating the problematic physical distance between communicative subjects in an age defined by late capitalism's requirement that bodies move fluidly across geographic space. Haptics accepts our mediatic situation but recognizes the epistemic, affective and communicative challenges that arise from the process of extending and abstracting bodies through audiovisual media. Haptic devices represent a sustained attempt to ameliorate the purportedly deleterious effects of mediated communication by mediating the sense of touch – by suturing another sense organ onto mediated sensorium. However, as with visual and audio media, mediation does not involve simple and transparent *extension*, but instead, an ideological and value-laden process of *transformation*. Haptic interface design, as a formalized interdisciplinary research field, forces a structured approach to touch that disaggregates, quantifies and operationalizes touch's variety of interrelated functions.[1] This has profound ramifications for the technological reconstitution of touch. As one specific application of haptics technology, the Hug Shirt embodies a division between the affective and informatics functions of touch, with the Hug Shirt intentionally designed to privilege the affective function of touch over its informatics counterpart. This division is expressed in the device's discursive construction, with CuteCircuit situating the Hug Shirt in its advertising and promotional literature as a therapeutic salve to the crisis of touch, as well as in its materiality, with the configuration of the shirt's sensors and actuators intended to capture and transmit a selected modality of interpersonal touching.

A CRISIS OF TOUCH?

New digital communication technologies, ostensibly intended to help ease the feelings of distance between remote communicative subjects, have recently come under criticism for driving a wedge between proximate subjects in situations of face-to-face communication. According to some recent commentators, including Richard Kearney in 'Losing our touch' (2014) and Paula Cocozza in 'No hugging: Are we living through a crisis of touch?' (2018), digital communication technologies are displacing physical contact, as we increasingly touch our devices and screens more than we touch each other. Consequently, we are faced with the challenge of being in constant communicative contact with those in our affective networks while missing the well-documented psychological benefits derived from the physical act of interpersonal touching.[2] Kearney crystalizes what he takes to be the challenge posed to tactility by the widespread practices of digitally mediated communication: 'our current technology is arguably exacerbating our carnal alienation. While offering us enormous freedoms of fantasy and encounter, *digital eros* may also be removing us further from the flesh' (Kearney 2014: n.pag.). The growing embrace of digital communication intensifies our transformation into a 'more and more a fleshless society' (Kearney 2014), marked by a loss of interpersonal, flesh-on-flesh touching. Touch in this formulation is something irreducible and untransmissible, screened out by the materiality of digital media interfaces.

Tiffany Field, a long-time proponent of increased, structured forms of nurturing touch and founder of Touch Research Institute at the University of Miami, similarly attributes the decline of touch to the increasing prominence of digital media. In a recent interview, she decried the lack of interpersonal contact in public spaces; in an airport 'not a soul was touching another. Even 2-year-olds were sitting in carriages with iPads on their laps' (Cocozza 2018). Field speculates that this growing lack of interpersonal touch, with children touching their devices more and each other less, is responsible for a host of negative effects, including a rise in the number of cases of irritable bowel syndrome and fibromyalgia and an increase in aggressive behaviour due to the growing absence of human touch in their lives. The problem, for Field, is not that these children are not touching, and it is that they aren't touching and being touched *by people*. Touching objects, then, ranks far below touching humans in Field's hierarchy of beneficial touches.[3]

In both Kearney and Field's narratives, distancing ourselves from human tactile contact distances us from our humanity (or, as Cocozza summarizes, 'without touch, humans become less, well, human') (2018: n.pag.). The way to regain this lost sense of humanity, then, is to reconnect with the sense of touch itself, to rediscover what it means to be human first by cultivating practices of interpersonal touch and second, by consciously attending to the significance of these flesh-on-flesh contacts. 'Full humanity', Kearney writes, 'requires the ability to sense and be sensed in turn' – a bidirectional touching that embraces the risk and reward inherent in the act of touch, different from the unidirectional model of touch as manipulation offered by the touchscreen interface (2014: n.pag.). Digital communication technologies function as barriers to touch and, by extension, barriers to our humanity.

Although virtual reality has long been hailed as a means of imbuing distanced communication with a sense of embodied intimacy (see, for example, Rheingold 1990), the arrival of a new wave of VR hardware in the past half-decade reinforces rather than challenges the audiovisuality of mediated communication, with the human haptic system largely left behind in the body's migration into virtual worlds. The hardware and software systems that extend the senses across space and time do so unevenly, with touch screened out by the materiality of communication technologies. Contrary to McLuhan's (1964) suggestion that electronic media are fundamentally tactile and will restore wholeness to the fragmented mediated sensorium, the rise of digital media seems to have exacerbated rather than abated this crisis, with communicative subjects shielded from each other via screens and absorbed in their devices, touching screens rather than touching each other. Here, I am not necessarily interested in evaluating the veracity of these narratives. Rather, examining these claims shows first, how they position digital technologies as casual agents in the decline of touch and, second, how they stage a technological response to this crisis, taken up in the next section, where new devices for remote touch attempt to ameliorate declines in flesh-on-flesh contact.

However, as much as it may be fashionable to blame digital communication for causing this crisis, its origins predate the rise of digital media. With the publication of his foundational *Touching: The Human Significance of the Skin*, humanist anthropologist Ashley Montagu identified what he understood as a pressing crisis in American culture, where infants and young children did not receive enough tactile contact from their parents. Montagu's book was the culmination of decades of research, a sustained investigation into the psychology of touch that began with his piece 'The sensory influences of the skin' in 1953. Throughout *Touching*, Montagu lamented a lack of attention to the skin in areas of inquiry ranging from poetry to psychology, pressing the case to actively revise our practices of interpersonal touch. 'As a sensory system', Montagu wrote, 'the skin is the most important organ system of the body' (1971: 8). To counter this passive neglect of touch, he offered the 'somatopsychic or centripetal approach', which emphasizes 'the manner in which tactile experience or its lack affects the development of behavior' (1971: x). The goal of Montagu's study was to identify 'the kind of skin stimulations necessary for the healthy development of the organism' and relatedly, the effects of 'insufficiency of particular kinds of skin stimulation' (1971: 12). The skin, Montagu concluded after an exhaustive rehearsal of evidence drawn from studies of both human and animal development, is a vital communicative space, used to send both positive messages (caressing, cuddling, holding and stroking) and negative ones (as with various forms of corporal punishment). Pointing to the general lack of nurturing touch in western cultures – and calling out the United States as a particularly flagrant offender – Montagu cautioned that 'inadequate tactile experience' at a young age would result in 'a consequent inability to relate to others in many fundamental ways' (1971: 319).[4]

Montagu's interest in touch coincided with Frank Geldard's sustained investigations, from the 1940s on into the skin's communicative capacities. Suggesting that modern communication technologies had overcrowded the visual and audio channels, Geldard began searching for vibrotactile communication systems that could bypass these channels, routing images, sounds and words through the skin. Montagu discussed Geldard's work frequently in *Touching*, with his formulation of 'the mind of the skin' (1971: x) echoing Geldard's claim that he was searching, in developing his various tactile languages, for 'the tongue of the skin'.[5] Both Geldard and Montagu positioned the skin as an important and neglected organ ('the largest of the body' as Montagu pointed out frequently), seeking to win converts to their cause of rescuing the skin from its ongoing neglect. However, their rescue strategies differed: Montagu advocated a return to and embrace of interpersonal nurturing touch, while Geldard designed machine systems that could productively mobilize the skin as a communicative agent. For Montagu, touch could only be salvaged through a return to the human; for Geldard, suturing new apparatuses onto the skin held the key to unleashing touch's power as a communicative agent. Considered together, their efforts illustrate the emergence of a cross-disciplinary valorization of touch, articulated in response to a purportedly deleterious crisis specific to western culture.

Touching sparked a sort of evangelism around touch that took shape in the subsequent decades, with Montagu's fellow travellers stressing the importance of 'bring[ing] touch back into modern society' (Field 2000: 151) and restoring intersubjective tactile communication, especially between parents and their children. A range of organizations signed onto this mission, including the Touch Research Institute Field helped found in 1992 and the Heart Touch Project (established in 1995). Their aims were decidedly activist and pragmatic, seeking to further knowledge about the therapeutic advantages of touch and 'reverse the minimal-touch problem that keeps us from fully experiencing its many benefits' (Field 2014: ix). While these efforts have progressively gained momentum over the past fifty years, part of what the medical historian Robert Jütte refers to as 'the new pleasure in the body' (2005: 238), Field's ongoing lament about the lack of interpersonal touch in public spaces suggests that the crisis Montagu identified is worsening, rather than abating.

For my purposes, the crucial question here involves specifying the role proponents of the crisis narrative feel digital technology has played in causing or intensifying this crisis. In an updated edition of her book *Touch* (2014), Field blamed teen mobile communication practices and online dating for an increased aversion to touch, implying that teen reliance on digital media was causing a problematic alienation from interpersonal tactile contact.[6] Likewise, Kearney seems nostalgic for some mythical predigital moment when our practices of touch were operating at an optimal state. Both versions of the story seem problematically romantic in suggesting a retreat to non-mediated forms of communication and troublingly normative in their implication that there is some universally correct amount of touch teens should be dosing themselves with in order to maintain their mental health.

Rather than speak of a crisis of touch, I would suggest instead that there are multiple crises of touch unfolding simultaneously and in response to each other. The crisis Montagu identified – the one Field continues to work to abate – is a crisis of *not enough* touching: a crisis defined by touch deficiency. However, this crisis exists in relationship to an ongoing crisis of *too much* touching, defined by an abundance of nonconsensual sexual touch. The #MeToo movement, which documented and called attention to rampant workplace sexual assault and harassment across virtually all sectors of labour, highlights the extensive problems that arise from people feeling too empowered to touch others. Moreover, unwanted touch often comes under the guise of beneficial affective social touch that Field suggests we need more of. Field, at times, seems to imply that we've overcorrected in response to this second crisis, noting that coaches of children's sports teams are often afraid to hug their players, concerned about the civil and criminal ramifications of misconstrued or unwanted touch. But this second crisis is a crisis of power as much as it is of touch, involving a right to bodily autonomy and socialization to norms of consent in relations of touching. From this perspective, a retreat from physical touching – a move to communication mediated by the shield of digital technology – can be understood not as an incidental touch deficiency but instead as a strategic response to the power relations encoded in tactile contacts that occur within the context of a patriarchal society. The interface then becomes a type of armouring, permitting communication with the body safely shielded from tactile contact. The question then becomes one of how to let some touches back in while maintaining the shielding of the body.

THE TECHNOLOGICAL RECUPERATION OF TOUCH

While the crisis of touch narrative predates the rise of digital media, the audiovisuality of the digital media interface arguably deepens this crisis, intensifying the cultural privileging of seeing and hearing over and against touch. As a positive response to this lack of touch in networked computer-mediated communication and human–computer interaction, an array of researchers working under the banner of haptics – including those in robotics, computer science, human–computer interaction, psychophysiology and neuroscience – have spent the last several decades attempting to develop technologies capable of transmitting touch through the same electronic networks that transmit images and sounds. Although transmitting touch provides a unifying goal for this research, the means of facilitating this transmission vary greatly, embodying strategic decisions about which modes of touching should be passed through digital communication networks. From complex devices like gloves, exoskeletons and bodysuits to simpler instantiations of haptics such as the vibration feedback motors found in video game controllers, pagers, smart phones and wearables, haptics technologies each transmit selective versions of touch. The material configuration of these devices is informed by their intended use: a glove might be designed so that it allows for the perception of object properties through active palpation, whereas a bodysuit would aim to provide a more comprehensive tactile image of a virtual environment by projecting sensations to the arms, torso and legs. Using vibration actuators in smart phones and wearables, by contrast, allows for the transmission of simple messages via coded patterns of vibration, shrugging off the goal of haptic realism often pursued in more robust systems. Doing haptic interface design, then, necessarily elevates some modes of touch over others. At the material level, the model of touch replicated by haptics technologies is marked by the values, goals and ideologies of the designer.

At the discursive level, the field of haptic interface design involves an ongoing revalorization, rediscovery and rearticulation of touch's centrality to human existence. In attempting to reconstruct touch through an electromechanical interface, haptics practitioners gain a deeper appreciation for touch's complexity as a neurological, physiological and psychological system. In justifying the value of haptics research and applications, hapticians rehearse the crisis of touch narrative, situating their individual research outcomes as part of a more general effort to recuperate and restore the lost sense of touch. Hapticians become evangelists for touch, situating touch's technological reconstruction via haptics as a solution to a contemporary crisis of touch that takes different forms depending on the problem a given interface is addressed to. In these formulations, the increasing use of digital media interfaces does not *necessarily* lead to the decline of touch. Rather, they are motivated by the belief that the successful engineering of haptic devices can facilitate a new appreciation for and empowerment of touch.

As the field has matured over the last 30 years, the labour of designing devices to transmit touch has become increasingly specialized, with new subdivisions each tackling dedicated aspects of the problem. The Hug Shirt falls under the banner of *affective haptics*, a subfield of haptics that uses digitally transmitted haptic cues to elicit a range of emotional responses in the receiver of the tactile message. Other devices in this long tradition include vibrating wristbands (such as the Bond Touch bracelet), actuator-enabled armbands (see Chang and Israr 2020), the Apple Watch's Taptic Engine (which, according to Apple's promotional materials, allows wearers to send 'something as personal as your heartbeat' to other wearers), handshake simulation gloves (see Jewitt et al. 2020, for a summary) and a spate of other vests and shirts intended to remotely transmit affective bodily contact.

The Hug Shirt was an early entrant into this field, comparable in aim and function to a range of devices developed subsequently to its debut in 2002, including the Huggy Pajama (Cheok 2010), iFeelIM! (Tsetserukou et al. 2009), The Hug (Gemperle 2003) and Hug Over a Distance (Muller et al. 2005).[7] Some of these devices, such as the Adrian David Cheok's Huggy Pajama, use more robust mechanisms of capture, storage and transmission to transfer hugs between communicative subjects, where others, like the Hug Shirt, sacrifice haptic complexity in the aspiration to commercial viability. Despite their material differences – some use inflatable pockets of air, and others employ vibration – they employ similar narrative frameworks to make the need for these devices seem urgent, pointing first to psychological studies that highlight touch's importance to human emotional development (such as those provided by Field and Montagu) and, second, describing how our contemporary communicative situation problematically leaves touch behind. Designers of hug shirts and other affective haptics applications, then, rehearse and intensify the crisis of touch narrative. The promotional materials for the Hug Shirt, for example, present the normative claim that 'people need to be touched at least 70 times a day!', urging the Hug Shirt's potential wearers to 'start noticing how many times you shake hands or hug a friend, and you will see that it really makes you feel good' (CuteCircuit 2008: n.pag.). Noting that 'an increasing mobility of humans throughout the globe, due to business or study reasons, has brought family members to spend most of their time apart from each other', the Hug Shirt is offered as a technological salve for a capitalist wound. It is not so much the our mediatic situation that makes the need for affective haptic urgent, but rather, the conditions of labour in contemporary society, which demand that bodies move fluidly around the global, that people spend time physically distant from their loved ones.

Communication technology, from this perspective, is not situated in opposition to the human, but instead, embodies an ideology of what Arvind Rajagopal (2020) calls 'communicationism' (357) – a belief with origins in the Cold War that communication through contemporary media technology enables human agency and potential. The problem of mediated communication, in this framework, is not media per se, but the incomplete and piecemeal way that they extend the human body. Any deficiency of communication technology can be solved by simply adding better technology. Playfully refusing the need for a nostalgic return to face-to-face communication, hug shirts are not explicitly *not* intended 'to replace human contact', instead providing a way 'to make you happy if you are away for business or other reasons and you miss your friends and loved ones' (CuteCircuit 2008: n.pag.). Crucially, communicationism, for Rajagopal, is an ideology that celebrates technologically mediated communication as a pathway to the expansion of human agency while obscuring the ways communication technology reconfigures human agency and conditions of techno-sociality, ceding control over communicative processes to machinic systems. Hug shirts and other affective haptics applications seek to use media technology to connect us back to a neglected sense of touch, drawing on mountains of psychological studies to make the case that the need for such a reconnection to touch is vital and urgent. But this series of rhetorical moves conceals two crucial transformations. The first involves the way that psychology, in its thorough quantification of touch's emotional benefits, has instrumentalized touch as a means of maintaining a sort of normative affective homeostasis.

Descriptive studies of the link between touch and emotion become prescriptive calls to maintain a specific regimen of touching and to turn to an assistive technology when that regimen cannot be otherwise maintained. The second involves the sleight-of-hand substitution of technological touch – the touchless embrace – for its flesh-on-flesh counterpart. Despite the strident claims by CuteCircuit and other hug shirt designers (Florian Muller is similarly insistent that his Hug Over a Distance 'will never replace a real hug') that their shirts are intended as a supplement rather than substitute for real hugs, situating their devices as a response to the crisis of touch suggests that they believe virtual hugs can accomplish at least *some of* the emotional work performed by their fleshy counterparts. Both this psychologization and this technologization of touch entail processes of rationalization, discipline and control, passing human touch through machinic processes of translation and articulation. The touch that emerges from these processes is something of an imperfect doppelganger, possessing markers of the original, but distinct enough in sensate appearance that it won't be conflated with its copy.

There is a temptation to link virtual hug technologies more generally to teledildonics the remote and computer-controlled stimulation of the sexual organs (see Machulis's contribution in this volume) and to cybersex – but here again, we can see how haptics operates by dividing the various tactile modalities. Despite their material similarities and genealogical intersections, teledildonics and affective haptics are cleaved off from one another. Unlike affective haptics, teledildonics has no formal institutional home; the field of haptics attempts to create and maintain a sort of firewall between technologies of touch and technologies of sexual touch. Papers on cybersex applications for haptics are almost universally absent from the major haptics conferences, whereas affective haptics has emerged as one of the major pillars of haptics research (the 2020 *IEEE Haptics Symposium*, for example, features a keynote session focused on the challenges of designing technologies that can facilitate the transmission of affective touch). The design of hug shirts follows this imperative, attempting to desexualize the technology and disavow any connection back to sexualized touching. In addition to being discursive and institutional, then, this segregation between sexual and affective touch is embodied in the design and material configuration of hug shirts, with the configuration of each shirt specifying the body parts that can be stimulated, along with the range of possible sensations that can be transmitted. The near impossibility of stimulating the entire surface of the skin, and the entire variety of skin senses, means that designers must make strategic decisions about how haptic devices replicate touch. The subdivision of touch according to its affective, sexual and informatic functions steers these design decisions.

Affective haptics orders touch, and the human emotional system more generally, as an optimizable series of inputs, embedding a normative model of proper touches into the design of these devices at the material level, while at the discursive level, affective haptics reifies the narrative framework that locates touch in a moment of crisis. It presents a response to the crisis of touch that embraces rather than runs from the technologization of the human sensorium.

THE SEMIOTICS OF A (REMOTE) HUG

To this point, I have approached the Hug Shirt in generalities, attempting to understand how it participates in a broader project of rehabilitating touch through technology. However, the success of this overarching project – which aims to ameliorate the deleterious effects of distanced interpersonal communication – hinges on the ability of individual devices to convincingly transmit emotion via networked haptics. It is therefore vital to unpack the techniques and mechanisms employed in the Hug Shirt, as a way of specifying the interdependence of the technical and social in the haptic transmission of affect. At the material level, the Hug Shirt uses a combination of strategically positioned sensors and actuators to capture and transmit sensations of hugging. The placement of these sensors and actuators was informed through an iterative design process that involved what Rosella and Genz called 'body storming sessions' where they 'asked people to hug each other for a very long time', mapping 'the position of their hands on the other person's body' (CuteCircuit 2020: n.pag.). Those points of most frequent contact provided a blueprint for the placement of the sensors on the Hug Shirt – a way to effectively compress the vast haptic data that comprises a hug down into a string of numerical values transmittable over a cellular network with limited bandwidth. Here, it is worth recalling that these virtual hugs move via SMS; network transfer speeds have long been an obstacle for transmitting touch (to the extent that many have predicted the deployment of 5G internet will bring about a revolution in distanced touch, referred to hopefully as the 'tactile internet'[8]) and the Hug Shirt successfully dodges these by creating a hug so narrowly compressed that it could pass through a telephone network. CuteCircuit frames this process of hug capture and compression in explicitly mediatic terms, explaining that the Hug Shirt 'records a hug like you would record a movie'. I have elsewhere referred to as the logic of analogue medialization (Parisi 2018): the notion that touch can become like the mediated senses of seeing and hearing by passing it through a technological filter. The logic of analogue medialization provides a heuristic for understanding touch technology that makes it intelligible by way of analogy while obscuring fundamental differences between touch's material underpinnings and the material process involved in seeing and hearing. The idea that a hug can be recorded like a movie situates the Hug Shirt as a camera that captures and stores a haptic recording of a hug (Katherine Kuchenbacher offers a similar analogy with her 'haptography' system; see Kuchenbecker et al. 2011). But touch is a sensory system distributed throughout the body; it is difficult to localize to a single site. And the haptic movie the Hug Shirt captures is decidedly low resolution: it only samples the vast data of touch, reducing the complex assemblages of haptic sensations a hug contains to a mere trace of the original, as captured by the Hug Shirt's sensors, stored as data and then replayed for the remote subject.

Here, we can think of the hugs not as *transmitted* but instead as *signified*: the hug produced by the shirt is not, as CuteCircuit points out, intended as a replacement for a real human hug. Instead, the Hug Shirt allows for the transmission of what CuteCircuit describes as a hug *signature*: a sequence of distributed tactile sensations relayed through the shirt that *signifies* the hugger for the hugged. Points of contact between hugger and hugged are rendered not as continuous tracts of squeezing but rather as discrete points of vibration – a poor resolution hug, consisting of only a small handful of contacts between the hugging arms and the hugged shirt, compressed to fit the affordances of information transmission networks. But a poor resolution hug *might be* good enough. The Hug Shirt, because it is only just now seeing a proper commercial release, presents us with the empirical challenge of determining what counts as a good enough system of haptic signification. How accurately does the Hug Shirt need to replicate the sensation of a hug for the virtualized hug to have meaning for and elicit emotion from the recipient of the digitized hug? CuteCircuit's claim that each person would have their own unique 'hug signature' to mark their hugs off as distinct from others is grounded in a psychophysical proposition: each hug could be objectively and quantitatively distinct while also being experientially similar and even indistinguishable. Psychophysics, as it was formulated in the mid-nineteenth century, initially concerned the capacity of experimental subjects to notice and not notice differences between sensory stimuli, a tacit recognition that human sense perception is fundamentally flawed, inadequate and inconsistent, in comparison to mechanical forms of measurement. The question then becomes not 'are two hugs distinct?' or 'can two hugs be perceived as distinct by a user in a heightened state of attention?' but instead 'will a particular individual wearing the Hug Shirt be able and willing to notice the differences between two distinct hugs?'. As CuteCircuit brings the Hug Shirt to market (in late 2019, they made it available for pre-order, but they have yet to begin shipping units), its commercial success or failure might depend on precisely this question: if the wearer can't differentiate between hugs coming from the range of senders in their affective network – a hug from a romantic partner, from a spouse, from a child, from a parent or from a friend – then the Hug Shirt may fail to deliver on its central promise of facilitating networked intimacy.

This psychophysical challenge, as media theorist Friedrich Kittler (2006) pointed out, underpins all technical media in some form or another. The .mp3 codec, for instance, depends on a structured knowledge of listeners' capacity to notice and not notice the difference between different audio frequencies, with this knowledge obtained through repeated studies of listeners' experience in controlled conditions of exposure (Sterne 2012). But with image and sound media, these questions have all been answered in some form or another, even if debates still exist around the ear's ability to distinguish, for instance, between the sound produced by vinyl and a lossless digital file. Most deployments of haptic media have been fairly limited in scope thus far, leaving us with a void of empirical data on how the collective social body will respond to virtual hugs and touchless embraces. It may very well be that, given the right device, we would accept a low-fidelity hug as an acceptable and effective proxy for the real thing. We would learn to interpret various arrangements of cacophonous vibrations erupting from different points on our shirts as distinct 'hug signatures' from our loved ones. The Hug Shirt, as Rosella suggests, only needs to index and point to our memory of a person's hug, rather than replicating it faithfully, the same way that a digital reconstruction of a loved one's voice by a smart phone successfully conjures an effective response, despite its imperfections.[9] We do not necessarily require a photorealism for touch – which has long provided an assumed endpoint for haptics research – to have meaningful and powerful experiences of touching and being touched from a distance. That question of the *good enough* for a particular medium will always be a moving target, with cultural responses and adjustments to new technologies in constant dialogue with the technical capacities of a given device.

Conversely, we can also speculate that potential wearers may simply be unwilling to put in the work required to distinguish between the range of possible hug signatures the Hug Shirt's cacophonous vibrations produce. Abandoning the goal of touch transparency in favour of a low-fidelity haptic image, in this scenario, would prove an impediment to the Hug Shirt's success, with the dermal image it projects onto the body marking the remote hugger's absence, rather than their presence. The Hug Shirt fails to realize – and in fact does not even seek to realize – the dream of touch transparency. The low-fidelity haptic image it projects onto the body cannot run from its artifice: rather than conjuring the loved one's physical presence through these tactile signifiers, it might accomplish the opposite, with the fundamentally broken haptic simulation reminding us instead of their absence. Here again, the cultural and the technological intersect: the sensations captured by the sender's shirt, of course, are only a small sampling of those the hugger themselves feels upon giving the hug – a remarkably poor copy of the original hug, missing more haptic data than it contains, absent temperature, pressure, tension and even spontaneity. The hug-that-is-not-a-hug can only ever perform the constraints of the device. This is not to echo Anne Cranny-Francis's humanist critique of the hug, which raises the concern 'that if human users come to accept a technological hug instead of a human touch, they will lose the specificity of "the human"' (2013: 157) but instead to speculate on some of the Hug Shirt's possible limits as an effective signification system.

The capacity to transmit an emotionally impactful and meaningful hug depends, then, on the Hug Shirt's wearer being able to distinguish between different configurations of machinic hugs: the wearer must learn what is it that marks the hug from one's spouse off from the hug from one's child, as it is relayed through the Hug Shirt. They must acclimate themselves to the capacities of the machine, learning to accept the touchless embrace as an emotionally impactful if imperfect signifier of affection.

EXPERIMENTS IN REMOTE TOUCHING

To close, we can pull back from this individual case study to raise the question of why, despite such a sustained research effort, hug shirts have failed to take hold in anything but our imagination. Part of the answer might paradoxically lie in the rapid global increase in commercial flights, with bodies now moving across the world at a previously unknown pace and volume. While this may result in our temporary distancing from those in our affective networks, mobile audio, visual and textual modes of communication function effectively to sustain these affective bonds until we can return. In his groundbreaking 1971 dissertation on computer-generated touch, Michael Noll suggested that future technologies of remote touch might reduce the need to travel, if they projected a robust version of touch across space. Noll presciently asked 'why transport people, with the resultant waste of energy and human time; why not transport signals instead?' (Noll 1972: 10). Already in the 1970s, nearly twenty years before the formal designation of Computer Haptics as a designated field of research, Noll anticipated the rapid advancement of remote touch, to the point where it would be a good enough substitute for tactile physical presence, similar to the way television and radio were already functioning as good enough extensions of the eyes and ears. Where touch is concerned, however, we have continued to prefer the transmission of bodies to the transmission of signals: for at least the past twenty years, it has been technically feasible to send something approximating a hug over the internet, with a host of devices in the tradition of affective haptics designed and brought to market in an effort to achieve this goal. Despite their ingenuity and innovation, and despite attracting consistent interest from the popular press, none of these devices has yet to achieve widespread adoption or commercial success. Remote touch has yet to be deemed – and may never be deemed – an acceptable substitute for flesh-on-flesh touching, in part because physical touch is so ready at hand.

The cultural impact of affective haptics, however, is not dependent solely on its commercial success. The Hug Shirt, regardless of its future trajectory, has already succeeded at challenging our ideas about touching, prompting us to reflect on the value both of touch and its digital surrogate. Alongside other developing technologies of networked touch, the Hug Shirt makes us attentive to the ways touch is being transformed by efforts to pass it through digital networks. This transformation is as much social as it is technological: we are currently undergoing an active renegotiation of what it means to touch, in response to our changing conditions of technosociality. As Carey Jewitt et al. suggest in an extended report on their In Touch: Digital Touch Communication project, 'the contemporary moment of digital touch innovation means the social norms for their use are un-developed and in flux' (2020: 59). From this perspective, the collection of odd gloves, bracelets, shirts and armbands in the dustbin of affective haptics is not an archive of failure, but rather, an attempt at bringing haptics into alignment with socially situated use cases. Likewise, the string of popular press articles stretched out over the past twenty years hailing the imminent ability of affective haptics to restore tactile intimacy to our networked communications should not be treated as inaccurate and sensationalist prognostication, but instead indicate a collective cultural desire to rematerialize tactility in response to the increasing dematerialization of communicative processes. This fantastical touchless embrace teases us with the promise of a (fashionable) restoration of touch, a perpetual and instant remote contact with our loved ones woven seamlessly into our daily lives. The Hug Shirt and other devices like it represent strategic responses to the challenges of intimacy in an age of digitally networked communication.

Positioning these devices as a solution to the ongoing crisis of touch situates the body as an affective machine that requires specific modes of tactile contact in order to function optimally and efficiently, part of a more general therapeutic ethos mobilized around touch. These devices embody normative assumptions about the affective function of touch, instrumentalizing touch and tactility as a means of maintaining mental and physical well-being. A regimen of (virtual) tactile contact, maintained through digital interfaces, aspires to become part of what Kaisu Hynna, Mari Lehto and Susanna Paasonen describe as the 'affective body politics' of digital media, involving 'the carnal ways in which bodies experience practices of governance' (2019: 1). Moreover, these haptic technologies bring with them new fragmentations of touch, new divisions of the tactile senses intended to facilitate the technological reproduction and reconstruction of touch from a distance. Their material configurations embody strategic decisions about which modes of touching should be transmitted and reproduced, encoding assumptions about how to properly maintain mental well-being in the design of the device. Affective networks, in these models, are incomplete and deficient without touch.

Throughout this piece, I have approached the Hug Shirt with a sustained scepticism, warranted, in my view, because the technology has seemed so tantalizingly close over the past two decades, promising to provide an experimental trial in the capacity of affective haptics to blend unobtrusively into the mediated sensorium. This scepticism, however, risks obscuring the sense of joyful enthusiasm and playfulness that Rosella and Genz bring to the design process: when I spoke with them in 2019, they described the Hug Shirt both as an experiment in distanced touch and as an expression of their philosophy that wearable technology can be comfortable and visually striking. The Hug Shirt does not ask its wearer to collapse the distance between hugging bodies but instead seems to playfully revel in its own artifice, celebrating rather than running from its perpetual novelty.

ADDENDUM: COVID-19 AND THE NEW CRISIS OF TOUCH

As I am finishing this piece in March 2020, countries all over the world moved to mandatory quarantine and social distancing techniques, in response to the rapid and deadly spread of the COVID-19 coronavirus. In real time, we are witnessing a massive, life-or-death renegotiation of social touching protocols: the question of who can touch whom and, under what conditions, has been rapidly transformed, as social distancing practices require that we remain at least six feet away from others, avoid gathering in groups and confine ourselves indoors as much as possible. Travel between countries and regions has come to a crashing halt, with bodies that normally flow freely through space suspended in place. Those stricken by the virus often die without the comforting touch of their loved ones, with their bodies transformed into sites of contagion. This prohibition on touching the infected is not extended equally to all: healthcare workers, frequently lacking suitable prophylactic equipment, are subjected to extreme risk in being required to administer care and comfort to the ill. Suddenly and with little warning, the physical proximity that had been regarded as a marker of emotional intimacy has been re-coded as a signifier of potential mortal danger not just to individuals but also to the social body more generally.

Those fortunate to be in the professional and managerial classes have in many instances been able to work remotely, encountering their coworkers through an audiovisual interface that abstracts communication from the body. Similarly, this videoconferencing technology allows many of us to remain connected to our loved ones, as we can see and hear them through our screens and speakers. One common refrain in writing on tactility touch occupies a curious status as simultaneously vital and neglected, taken for granted as an assumed constant of human experience while being underappreciated and experienced without critical reflection. But when we suddenly stop touching, we are confronted, in its violent absence, by the extent of its importance in creating, maintaining and strengthening affective bonds. If 'social touch can promote trust and cooperation' as neuroscientist David Linden (2015: 21) notes, and if 'no real community endures without touch' as the communication theorist John Durham Peters (1999: 269) argues, what will happen as those bonds age without being reinforced? As with the older crises, the new crisis of touch, then, is simultaneously a crisis of too much touch and not enough touch: touch – even self-touch – must be regarded as dangerous and contaminating, but in necessarily withdrawing from it suddenly and en masse, we risk undercutting the social bonds that maintain our affective networks. The empirical question of 'how necessary is touch for meaningful interpersonal communication?' is now being tested, as communication moves almost exclusively online. Already, we are seeing a growing cultural appreciation of touch; in a *New York Times* graphic editorial, Kristen Radtke suggested that our social distancing comes at the cost of the tactile relationships that are constitutive of our humanity, citing

Field's claims about the importance of touch to physical and psychological well-being.

Crucially, the interfaces we're using to stay in contact remotely
screen out touch; while our eyes and ears may be relatively seamlessly
transported across space, the various components of our haptic system
remain stubbornly immobile. Despite decades of research, despite the
aforementioned coherence of haptic human–computer interfacing as
a dedicated field of investigation and engineering, human–computer
interfaces have been unable to provide touch with a remote capacity;
touch, through some intractable quality, seems to resist our efforts to
mediate it.[10] At least for the moment, from a practical standpoint, touch
exceeds and evades the existing capacity of mediation systems.[11] As
we sit cloistered in our homes, connected by pictures and sounds but
unable to feel the faces on the screen, as we watch our loved grow
sick and pass away, unable to provide them with comforting touch in
their last moments, sending a hug remains an impossible, elusive and
increasingly seductive dream.

References

Anon. (2008), 'Hug Shirt', in A. Steffen (ed.), *Worldchanging: A User's Guide for the 21st Century*, New York: Abrams Books, pp. 97–99.

Chang, Xi Laura and Israr, Ali (2020), 'Communicating Socio-Emotional Sentiment through Haptic Messages', *IEEE Haptics Symposium*, https://research.fb.com/wp-content/uploads/2020/02/Communicating-Socio-Emotional-Sentiment-Through-Haptic-Messages.pdf?. Accessed 1 March 2020.

Cheok, Adrian David (2010), *Art and Technology of Entertainment Computing and Communication*, London: Springer-Verlag London.

Cholewiak, Roger (2006), 'The Cutaneous Communication Laboratory at Princeton 1962–2004', http://tactileresearch.org/pucclabs/index.html. Accessed 2 March 2020.

CNN (2014), 'Hug it out, from a thousand miles away', CNN, 19 February, https://www.cnn.com/videos/living/2014/02/19/natpkg-orig-shirt-that-hugs-cutecircuit-nyfw-ancil.cnn. Accessed 2 March 2020.

Cocozza, Paula (2018), 'No hugging: are we living through a crisis of touch?', *The Guardian*, 7 March, https://www.theguardian.com/society/2018/mar/07/crisis-touch-hugging-mental-health-strokes-cuddles. Accessed 2 March 2020.

Cranny-Francis, Ann (2013), *Technology and Touch: The Biopolitics of Emerging Technologies*, New York: Palgrave.

CuteCircuit (2008), 'The hug shirt', https://web.archive.org/web/20080531185659/http://www.cutecircuit.com/projects/wearables/thehugshirt/. Accessed 2 March 2020.

CuteCircuit, (2020), 'The hug shirt: Reach out, connect, touch, feel,' https://cutecircuit.com/hugshirt/. Accessed 13 May 2021.

Field, Tiffany (2000), 'Obituary: Ashley Montagu', *Journal for Body Work and Movement Therapies*, 4:2, p. 151.

Field, Tiffany ([2001] 2014), *Touch*, Cambridge: MIT Press.

Gemperle, Francine, DiSalvo, Carl, Forlizzi, Jodi and Yonkers, Willy (2003), 'The Hug: A new form for communication', in *Proceedings of the 2003 Conference on Designing for User Experiences (DUX '03)*, New York: Association for Computing Machinery, pp. 1–4.

Howes, David (2018), 'The Skinscape. Reflections on the dermalogical turn', *Body & Society*, 24:1&2, pp. 225–39.

Hynna, Kaisu, Lehto, Mari and Paasonen, Susanna (2019), 'Affective body politics of social media', *Social Media + Society*, 5:4, pp. 1–5.

Jewitt, Carey, Price, Sarah, Leder Mackley, Kerstin, Yiannoutso, Nikoleta and Atkinson, Douglas (2020), *Interdisciplinary Insights for Digital Touch Communication*, Cham: Springer International Publishing.

Jütte, Robert (2005), *A History of the Senses from Antiquity to Cyberspace*, Malden, MA: Polity.

Kearney, Richard (2014), 'Losing our touch', *New York Times*, 30 August, https://opinionator.blogs.nytimes.com/2014/08/30/losing-our-touch/. Accessed 2 March 2020.

Kittler, Friedrich (2006), 'Thinking colours and/or machines', *Theory, Culture & Society*, 23:708, pp. 39–50.

Kooser, Amanda (2012), 'Facebook-connected vest hugs you when you get a "Like"', *Cnet*, 4 October, https://www.cnet.com/news/facebook-connected-vest-hugs-you-when-you-get-a-like/. Accessed 2 March 2020.

Kuchenbecker, Katherine, Romano, Joseph and McMahan, William (2011), 'Haptography: Capturing and recreating the rich feel of real surfaces', *14th International Symposium on Robotics Research*, Lucerne, Switzerland, 31 August–3 September, http://dx.doi.org/10.1007/978-3-642-19457-315.

10.1007/978-3-642-19457-315. Accessed 2 March 2020.

Linden, David (2015), *Touch: The Science of Hand, Heart, and Mind*, New York: Penguin Press.

McLuhan, Marshall (1964), *Understanding Media: The Extensions of Man*, Cambridge: MIT Press.

Montagu, Ashley (1971), *Touching: The Human Significance of the Skin*, New York: Harper & Row.

Muller, Florian Floyd, Vetere, Frank, Gibbs, Martiin, Kjeldskov, Jesper, Pedell, Sonja and Howard, Steve (2005), 'Hug over a distance', *CHI '05 Extended Abstracts on Human Factors in Computing Systems*, pp. 1673–976.

Noll, A. Michael (1972), 'Tactile man-Machine communication', *Journal for the Society of Information Display*, 1:2, pp. 21–29.

Parisi, David (2011), 'Tactile modernity: On the rationalization of touch in the nineteenth century', in C. Colligan and M. Linley (eds), *Media, Technology, and Literature in the Nineteenth Century: Image, Sound, Touch, Burlington*, VT: Ashgate Press, pp. 189–213.

Parisi, David (2018), *Archaeologies of Touch: Interfacing with Haptics from Electricity to Computing*, Minneapolis: University of Minnesota Press.

Peters, John Durham (1999), *Speaking into the Air: A History of the Idea of Communication*, Chicago: University of Chicago Press.

Radtke, Kristen (2020), 'Op-Art: What do we lose when we stop touching each other?' *New York Times*, 20 March, https://www.nytimes.com/2020/03/19/opinion/coronavirus-touching.html. Accessed 2 March 2020.

Rajagopal, Arvind (2020), 'Communicationism: Cold war humanism', *Critical Inquiry*, 2:46, pp. 353–80.

Rheingold, Howard (1990), *Virtual Reality*, New York: Simon & Schuster.

Simsek, Meryem, Fettweis, Gerhard and I, Chih-Lin (eds) (2019), 'The tactile internet', *Proceedings of the IEEE*, 107:2, n.pag.

Sterne, Johnathan (2012), *MP3: The Meaning of a Format*, Durham: Duke University Press.

Tsetserukou, Dzmitry, Neviarouskaya, Alena, Prendinger, Helmut, Kawakami, Naoki and Tachi, Samsu (2009), 'Affective haptics in emotional communication', *3rd International Conference on Affective Computing and Intelligent Interaction and Workshops*, Amsterdam, 10–12 September, pp. 1–6.

Turkle, Sherry (2009), *Alone Together: Why We Expect More from Technology and Less from Each Other*, New York: Basic Books.

Notes

1. As I have argued elsewhere, this fragmentation of touch into discrete, isolatable components did not originate with haptic interface design; instead, its roots are in the emergence of haptics as field of experimental psychology in the late nineteenth century. See Parisi, 'Tactile modernity: On the rationalization of touch in the nineteenth century' (2011).

2. A range of psychological studies have discussed, for instance, the ways physical contact results in a release of beneficial chemicals in the body that can help maintain a positive mood and boost immunity. Ashley Montagu's work, discussed below, was a flashpoint for these studies, which have grown both in number and sophistication since the 1970s.

3. This emphasis on the importance of interpersonal touch contrasts with other perspectives on cultivating touch. Notably, the Italian educator Maria Montessori sought to educate and refine her students' sense of tactile discrimination using 'tactile tables'.

4. For a pushback on Montagu's centrality in accounts of touch and the skin, see Howes (2018).

5. While teasing out these connections is beyond the scope of this chapter, Cute-Circuit recently found a new application for the Hug Shirt, adding more actuators and using it not to transmit hugs, but instead to transcode sounds for touch, in an attempt to allow the deaf to experience music through their skin. The new device, the Sound Shirt, departs from the tradition of affective haptics, intersecting instead with a genealogy of sensory substitution systems that route data for eyes and ears through touch. Geldard had a similar aim in founding the Cutaneous Communication Laboratory at Princeton (active from 1962–2004, long after Geldard's passing in 1984), which produced hundreds of scientific and technical papers largely aimed at 'the development of sensory aids for deaf, blind, and deaf-blind persons' (Cutaneous Communication Lab 2004).

6. Sherry Turkle (2012) offers a similarly alarmist evaluation of the disembodied communication that occurs through social media, suggesting that teens want a return to touch and tactility in their communicative practices.

7. See Cheok (2010) for a comprehensive list of devices.

8. See *Proceedings of the IEEE* special issue on the tactile internet for a range of perspectives (Simsek, Fettweis, and I, eds 2019).

9. Here again, this is the result of a cultural adjustment to technological constraints: AT&T's 'Reach Out and Touch Someone' ads for their long-distance telephone service were an attempt to convince consumers that they could have emotionally impactful exchanges with their loved ones mediated by the telephone.

10. Peters argues stridently that 'out of all the senses, touch is the most resistant to being made into a medium of recording or transmission,' noting that attempts to transport touch 'always fell eerily short' (1999: 269–70). I have taken issue with Peters on this claim in the past, noting, for instance, the widespread success of vibration feedback in video games.

11. Some systems – such as the Haptx Glove and Shadow Robot telepresence system – have been able to successfully project the sense of touch across space. But these systems, primarily intended for industrial applications, remain prohibitively expensive and limited in their deployment.

9

Sounds of Feeling

Currently, our intimate, emotional and emphatic relations as well as affective states are increasingly framed and shaped by interaction through interfaces, data-fication and algorithmic protocols. Tele-presence technologies extend our bodies beyond biological boundaries in time and space, trying to enable direct intimate reciprocal touch, such as kissing, and transmit the emotional charge inscribed in our physical, intimate interactions. But can an intimate kiss be translated into neuro feedback data? Can we measure a kiss and what kissers feel together? How does your kiss feel when translated into EEG format? Can we kiss online or transfer a kiss with its erotic feel? Would we want to save our private kisses in transparent databases to be used by others? And finally, how does a kiss sound and what is it like to hear it?

EEG KISS by Karen Lancel and Hermen Maat (2014–present) is a science-based artistic project oriented to investigate the potential of translating a kiss into neuro-feedback data pressed in sound. The experiment invites couples to the performance space in which they kiss with EEG headsets on to let the data of the act get transmitted by multi-sensory Brain Computer Interfaces and saved in a database. The project deconstructs the kiss to reorchestrate a new synesthetic kissing ritual. Through hearing, seeing and touching, all participants together are invited to share a kiss, in a communal neuro-feedback system for kissing. The outcome is a shared neuro-feedback system for *networked kissing*.

The project embraces a number of performances and live kissing experiments ('Digital Synaesthetic EEG Kiss' [2016], 'Kissing Data' [2018] and 'Intimate Data Symphony' [2019]) that measure kissers' brainwaves and visualize them into EEG data. A floor projection encircles the kissers with the real-time EEG data streaming as 'Dancing data' of communal patterns and flow in an immersive data scape. Each unique EEG Kiss audio-visual data sequence goes to a database on a dedicated website where it is available for online audiences.

In this publically shared form of kissing, the familiar sensory relations between 'who you kiss and who is being kissed, what you see and what you hear' are disrupted for an unfamiliar sensory synthesis, a 'digital synaesthetic' orchestration. The audification of a kiss through BCI measurement intensifies the actors and spectators' experience of sharing feedback and immersion. The combination of audification and spatial BCI data visualization enhances focus, concentration, immersion and feeling of safety. Moreover, 'seeing to be kissed' builds on neurological insights of mirror-touch synaesthesia in which 'seeing to be touched' is considered to evoke empathy. This marks a new frontier of the perception of senses and informs the new dimensions of organicity. Extended by technology, our bodies release new sensory affordances. This inspires new affective states that allow for a more comprehensive approach to human emotions, especially in relation to intimacy and human intimate performance.

Digital Synaesthetic E.E.G. Kiss

KAREN LANCEL AND HERMEN MAAT

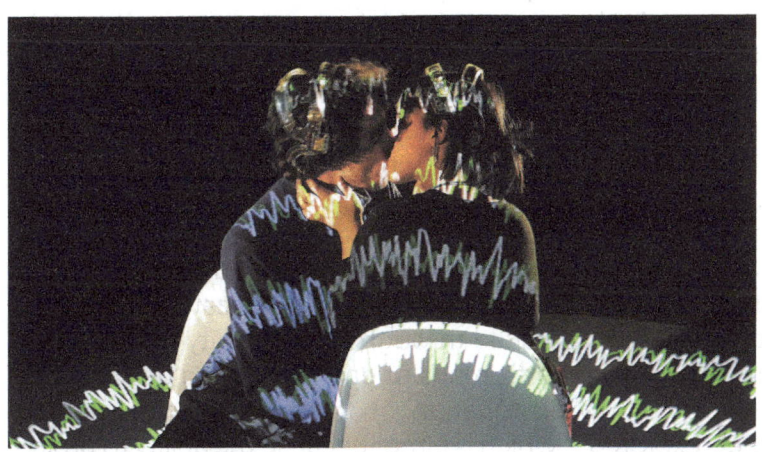

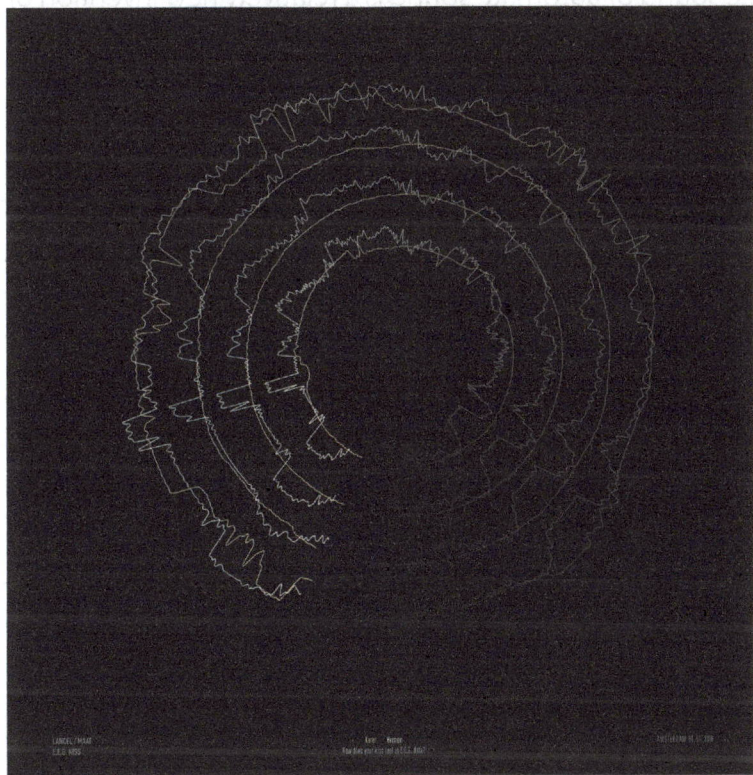

FIGURE 9.1: Lancel/Maat, *Digital Synaesthetic E.E.G. Kiss*, 2016. Audiovisual performative installation. Courtesy of the artists (photograph by Lancel/Maat) © Lancel/Maat.

FIGURE 9.2: Lancel/Maat, *E.E.G. Kiss Portrait*, 2018. Print. Courtesy of the artists (photograph by Lancel/Maat). © Lancel/Maat.

I Can Hear Your Feelings

ANDREW BLANTON

Sound is one of our most intimate senses. It is also the first sense we experience as human beings and perhaps the last. '*I hear therefore I am* [...] acknowledge(s) the way in which sound creates subjectivity through its own surplus as much as absence' (Pettman 2017: 1, original emphasis). An omnipresent experience, sound is what we feel in and with our entire body, through touch and emotional interpretation. In our daily lives, we are constantly moving through an infinite number of sounds. Our atmosphere vibrates to waves of sound washing over us, examined and transmitted by our ears, as vibrations are converted to electrical signal for cognition. Interestingly, sound remains under-developed as a medium for interpreting data. Visualization of data has a long history and can be traced throughout humanity whether it be in the form of early star charts, maps or data-recording devices such as the Quipu of the Inca. Sonification, on the other hand, is relatively new and underexplored – which is either because the ability to proce-durally make sound or automate sound is a relatively new invention or because we are a visually dominant species.

Sonification of the human experience can create profound and meaningful understandings of our emotional, lived experiences. Sound is an important way to interpret the world. The combination of multi-ple forms of sensorium creates our relationship with reality, and the combination of sonification with visualization offers new opportuni-ties for humans to better understand our environment, including the organic environment of their own bodies, especially their cognitive and sentient realms. Emotions entail particularly complex processes and often escape complete explanation. Sound has the ability to interpret emotions in ways that are felt rather than described. With advance-ments in EEG technology and increased accessibility to what was previ-ously very expensive and rare equipment, artist can now access sounds that have always existed but were never before possible to hear. We are developing new methods for mapping our aural landscape, trans-forming our ideas about sound and the ways we listen to the world. Just as we can see how hearing has changed with the development of technology, we can also look at the edge of technological devel-opment as an exploration into new ways of hearing the world. And as John Berger looked at new ways of seeing through the interpretation of fine art and advertisement, this new way of hearing can help us better understand our species-specific biases, and it can give us new ways to sense our psychology, physiology and ultimately help us gain insight into previously indescribable aspects of our own humanity.

Within the last forty years, there has been considerable growth in tools that enable the sonification of data. Beginning with MIDI as a protocol and then with the development of network-based technologies like OSC and ODOT, we see protocols and languages that are built specifically for the interpretation of data at audio rate. This in turn has opened new pathways for artist to create and build works that can give us deeper and more meaningful experiences with data and the interpretation of life. Early examples of working with the sound of data as art include Charles Dodge recording of the Earth's Magnetic Field, Larry Austin's sonification of the Canadian Coastline, The Hub, networked music ensemble and the proliferation of laptop orchestras in academic music programs. But sound has also been influential in the hard sciences as well. From the Gigercounter to the stethoscope to non-invasive forms of probing such as ultrasound and heart rate monitors, listening to our bodies has become an integral part of medical science. Sound is immersive; our hearing is omnidirectional as opposed to our visual sense, which is directional and forward looking. For this reason, sound can be atmospheric; we are enveloped by the experience. Sound offers different and new possibilities for interaction with data.

There is also a deeper emotional resonance with sound. Sound is foundational to cinematic experiences. Music has been said to be the language of emotion, and sound is an important component of our emotional interpretation, whether that be in the form of a movie or an art experience. The artwork *EEG KISS* by Karen Lancel and Hermen Maat hybridizes these worlds of presentation and representation allowing the audience to listen to emotions.

EEG KISS

EEG KISS explores the sonification of intimacy by asking the participant to augment one of the most intimate of human experiences: kissing. A performance and scientific experiment in one *EEG KISS* involves two participants positioned next to each other in closely placed chairs and wearing EEG headsets to record the electromagnetic output of their neurons. As the participants kiss, electrical signals flowing from their brains are interpreted by the headsets and go into a custom piece of software that converts the electrical signal into both sound and visuals. In that moment, their brainwave data is either projected on the floor or displayed on a nearby monitor and turned into sound, exposing the data of intimacy in a public performance setting. The process creates a circuit in which the act of kissing flows through technology between the participants and the audience. As Lancel and Maat explain:

> A soundscape is generated by the Brain Computer Interface, which translates the real time E.E.G. data of 'kissing brains' into an algorithm for a music score: an E.E.G. KISS Symphony. The public around is part of the kiss, of the sound and of the immersive E.E.G. data-visualization; both as an aesthetic experience as well as based on acts of kissing resonating when mirrored in their brains. Also, spectators wear headsets, to include their brain activity of responding mirror-neurons and imagination, of seeing and hearing a kiss, in the soundscape feedback system. People kissing and spectators co-create soundscapes and data-visualizations for a shared, immersive reflexive data scape. Every sound is unique. Each sound scapes is saved in a database from which all participants can download each other's 'kissing sound portraits'. (2019: n.pag.)

EEG KISS is an experiment in augmenting intimacy, in which a computer quantifies a deeply personal experience. As an attempt to translate the intangible effects of erotic stimulation into hard data, it translates the abstract into the concrete. The sounds of the EEG Kiss work like a stethoscope, listening to the inside of the brain, allowing the audience to hear the reactions of the kissers. Precise as it tries to be, this digital interpretation can never fully capture that moment. What it can, however, is to form an imprint of the moment, creating an association between the data and the action that strengthens the memory of the event while augmenting the lived experience of a kiss. Through this process of augmentation, we can reveal both characteristics of the kiss and the characteristics of the recording platform. The system gives an insight into the electrical signal of a kiss, producing an artwork that becomes an ephemeral space between the kissing couple and the audience. Connection that emerges in the process give rise to new intimacy as the artwork lives in the minds of the viewers. The artists admit:

> Important to our work is that experience of intimacy is, in fact, publically embedded. Our orchestrations are presented in (semi-) public spaces, to provoke reflection and dialogue on our changing perception of intimacy in entanglement with (non-)human others. We aim to facilitate participants to explore such experiences together. As hosts, we engage dialogue with participants

about their experiences, based on personal sense-making and tacit knowledge, social relations and environment. Secondly, dialogue with participants enables rethinking parameters for future concepts of AE neural networks for intimacy and empathy. Our orchestrations aim to encourage the design of a sensitive public space for shared intimacy, based on responsibility for the power of synchronizing through touching, watching, kissing, sharing presence and sound. Based on our work and fundamental research, for future concepts of Artificial Intelligence and Artificial Emotions, we currently create a social-sensory model at the Technical University of Delft (Participatory Systems Initiative). This model is meant for practitioners in the arts, design, architecture and city planning, academic research and science. For this model, we make use of artistic and scientific insights and of our artistic observations of participants interaction in our orchestrations in public spaces worldwide. (2019: n.pag.)

Conceptually, the work offers a number of metaphors for societal interactions, documenting and recording intimacy between two people. But ultimately, the work exists in the thoughts and emotions felt in the space, in the sounds of neurons firing and the lines drawn showing the electromagnetic pulses of the brain. This experience comes together to create an impression of a kiss but produces a unique data portrait of contexts – personal, private, public and cultural – that the kiss exists in.

EEG Kiss interacts with the universal meaning of a kiss. Being a fundamentally biological act, kissing bears many cultural meanings. It traditionally signifies emotional engagement, romantic inclination and sexual interest and has been considered 'one of the most effective bond-mediating courtship behaviour' with a potential for mating (Wlodarski and Dunbar 2013: 1416). Neurobiological studies of kissing show an intense hormonal dynamic of the act; it is manifested through the stimulation of olfactory semiochemical pathways running on androstenone and androstenol – substances responsible for sexual arousal and bonding. Due to this erotic function, kissing entails spatial intimacy and has forever been depicted as an act hidden from the public eye. It carries a strong sense of exclusivity (two people or so) and a reliance on the organic and psychological instincts of the engaged parties (lips only). Transmitted technologically into the environment of an experiment or an exhibition space, the kiss loses this intimate context by engaging various media, devices and apparatuses; with this, it develops new signification that may often contradict the imagery it traditionally entails. At least, that has been the experience of EEG

Kiss, which, as the artists explain, deconstruct the notion of privacy, emotional vulnerability and emphatic interactive-ness:

> In digital performance art in urban public spaces, unfamiliar, unpredictable haptic and somatic encounters, in symbiosis with technology, are often staged in playful, and ambivalent ways, to provoke immersive engagement and sense-making. Works such as *Body Movies* by Rafael Lozano-Hemmer and *Can You See Me Now?* by Blast Theory, seduced participants to appropriate and synthesize familiar and un-familiar biofeedback of visual and haptic connections. New syntheses are then based on combined processes of mirroring in digital representations. In this context, *EEG KISS* explores whether an intimate kiss can be shared by plural (more than two) participants when supported by a multi brain BCI orchestration. Firstly, we explored whether the tactile experience of kissing can be transferred to perception of seeing and hearing a kiss. The orchestrations show that for shared engagement in a kiss, the act of kissing needs to be re-orchestrated into multiple ambivalent, direct and disrupted sensorial connections and connections between physical presence and virtual, spontaneously emerging BCI representations. Secondly, the social aspects of intimacy need to be attended. Often, in artistic (telematic) sensory syntheses, public intimate social engagement is aroused through corporal, somatic vulnerability of an artists' body. In such orchestrations, participants are challenged to consider approaching, touching or even physically abusing the artist/performer, leading to new social, reflective connections. Vulnerability is core to these artworks and can be considered to be a feature of interdependency in social bonds, in relation social values of responsibility, witnessing, empathy and trust. Unfamiliar and unpredictable forms of vulnerability and responsibility often call for dialogue, to re-negotiate and reflect on these social values, in a public process of shared and personal sense-making. In *EEG KISS*, instead of an artists' vulnerable body, members of the public space are *in interplay with each other,* which requires a different approach. We call this approach 'Distributed Vulnerability'. In this approach, conditional to intimate experience of kissing being translated in sound in public space, is co-presence, witnessing and sense making with all participants. All participants together are invited to take interdependent roles of kissing (vulnerable) or of spectators (responsive witnesses). Analyses of these orchestrations show that shared social touch of kissing must evoke imagination

and a personal form of sense-making, as a vital component of emphatic response to touch, mediated by a host. (2019: n.pag.)

SONIC TECHNOLOGIES OF INTIMACY: A CLOSER LOOK

Technological augmentation of human organic functions is inscribed in the civilizational strife for effectiveness. We build tools, develop technology and extend our bodies to enhance our abilities while allowing for more productivity with less effort. We also get a perspective on many hidden aspects of our physical abilities that our bodies invariably poses, but which we cannot fully access from the confinement of our physical shell. Information that comes from technological extensions inspires new behaviours that often lead to technologies and experiments aimed at stimulating further cognitive and sentient advancement and imagery.

Sonification is technologically focused on highlighting and framing data that communicate on a more emotional, experiential and felt level. Certain works do this well, from Mahmoud Hashemi and Stephen LaPorte's *Listen to Wikipedia*, to John Luther Adams's *The Place Where You Go to Listen*, to Brian Houses's *Quotidian Record*, or *1945–1998* by Isao Hashimoto. A lot 'affective' sound-oriented technology is also used in medicine. For instance, ultrasonographic studies of Takotsubo syndrome offer sonic interpretations of a broken heart to datafy the symptoms and rehearse therapy. The studies explore the records of breathing patterns and heart rhythm, supporting sound data with electrocardiogram picturing. Whereas the organic sample in such studies seems accurate enough to provide a reliable input (the patients selected for such studies are psychologically and medically verified for a heartbreak), it is difficult to assess the veracity of the outcome itself. In other words, we have no certainty that the sonic imaging of emotions reflects their actuality. Synthesizers, in the most conventional sense, take a consistent stream of electricity, modulate that stream through amplification, attenuation and harmonic manipulation to vibrate a speaker cone. The whole process is a direct manipulation of electrical signal flowing out of a wall. In the case of sonification, we still have the initial electrical signal and the synthesizer modulating that electricity, except we add a data source and a source of automation that is controlling the synthesizer. And while we can understand this process from a mechanical perspective, the question remains, why and how this provides a more meaningful interpretation of an experience.

Experiments like *EEG KISS* operate on a feedback loop from the technologically transmitted input (the kiss) and the digital transformed output. In this sonified feedback system, the activity of the 'kissing brains' affects the BCI input manipulating the algorithm, after which, in turn, the brain activity itself is affected by the BCI audio output (the sound). As a consequence, in fact input and output inform a loop. This feedback loop is further informed by spectators' brain activity and social feedback in relation to the vulnerability and self-disclosure of people kissing. The algorithm that generates the sound patterns of each kiss – 'ticking and crackling' sounds (by electric disturbances of the 50-Hz system), 'water bubbles tickling' sounds (based on low tones) and bells tingling sounds (achieved through soft high tones) – makes use of pre-defined combinations and averages of both of the participants' EEG data signals. Although EEG systems measure mostly muscle movement, they are also part of reciprocal intimate processes, including motoric intention in the brain. As a consequence, the data signals predominantly emerge from motor intention and body move-ment of kissing. The sound adapts to various performative phases of the kissing process, with different sound patterns, separated manually by the host. A 'sound flow' is acquired by crossfading separate pattern 'spheres', based on artistic choices.

The participants of *EEG KISS* experiment have given a positive testimony to the experience of the shared data-sonification. They believe it has strengthened their connection and enriched the rela-tionship dynamic. Some admit that the kiss felt like it was being borne from the music. Others state that the sound made the kiss more intense and more focused. But not all interactions can be expressed in numbers. Sonic and visual representation can give a new insight into what emotions are, but their picturing does not reflect the entirety of emotional processes or biological reactions taking place in our brains (e.g. when the cocktail of dopamine, oxytocin and serotonin is being released at kissing). The creators of *EEG KISS* admit that such a reflection was never part of their intention. Neither was the mimicry of kissing.

> The multi-sensory design does not aim to imitate a kiss. Instead, the digital syneasthetic disruption and re-orchestration of senses aims to evoke both a new experience of kissing as well as dialogue and shared sense making on intimacy. The kiss was translated into real-time amplified data-sonification, to strengthen feelings of shared immersion and intimate connection between people kissing and spectators. The sound is part of a multi-sensory feed-back loop, essential to real-time sharing the act of kissing. While

in the orchestration, people kissing primarily share only *touch* (most often with eyes closed) and spectators can only *see* (kissing and data-visualization), sound facilitates shared experience in real time. The shared data-sonification has shown to increase participants' feeling of safety and involvement, both in time and in intimate connection with each other. Participants express for example: 'The sound makes the space reflective', 'This feels like a kind of sharing a trance' and 'This situation is weird but feels strangely safe'. Some stay for hours, talking quietly with each other. When sound is included, spectators encourage each other to kiss and the duration of kissing is longer. People of all ages, genders and cultures kiss, even strangers kiss. Most participants describe their experiences as being 'very intense'. Afterwards, other forms of individual sensory perception (of seeing and kissing) are related to these experiences of shared sound. (2019: n.pag.)

Translating emotions with technology is a rich world for exploration in art. There has been considerable work done in the field of data visualization around representation; we can see some good examples in the work of Edward Tufte's Envisioning Information, Steele and Llinsky's Beautiful Visualization or in more abstract terms in Form + Code. Just as augmented reality provides a layer of abstraction over our world in the form of a digital overlay of information to give us deeper understandings about the world around us, the augmentation of a kiss between two people attempts to derive further information through abstraction and interpretation of the brain's physiological processes that we call emotions.

SONIC DATA FOR NEW MEDIA INTIMACY

Emotional data is an important part of the online market and the future of social platforms. As our lives move to the internet, the interest in technological transmission of physical contact becomes a priority of a vast sector of the tech industry. Inventions such as Teletongue or Kissinger (prosthetics used for the transmission of mouth and tongue interactions at a distance) show the directions of that interest, informing that intimate data – like a digitally registered kiss or touch – might become an element of a future chat-up or foreplay. Serving as an attempt to quantify human intimacy, *EEG KISS* raises important questions into how data is being used in our lives to mediate reality and shape our understanding of human relationships. It also initiates dilemmas over contemporary networked reality and the continual push to gain more resolution in the augmentation with the technology of our everyday lives.

We are clearly facing an important design and interface questions with the adoption of social media and the quantification of our personal lives. Data is abstract, and most users do not understand what is being collected and how it is being used. Facebook, for example, works exhaustively to create data profiles that extend far beyond the data provided by users of facebook.com. So do many online dating platforms. The protection of data is an enigmatic area beyond the control of the users. This can be seen in the 2016 FitBit data breach that led to the exposure of 150 million customers' most intimate data including potential sexual habits. This can also be seen in the manipulation of fashion trends as well as elections through social media as demonstrated by the researcher and whistleblower Christopher Wylie from Cambridge Analytica.

That does not, however, prevent us from investing into the data reality. On the contrary, the advancing possibilities around the technologization of experiences invite (if not generate) new solutions for datafying or digitally capturing the least tangible aspects of our life. There is a beauty in using a kiss to explore the landscape of feelings and chemicals we call our emotions. There is also a beauty in having it 'engraved in sound', preferably in a shareable format that allows for exchanging it or storing it like a keepsake. Our experiences are not 'solid' and can never truly be re-experienced past the moment in which they happen. Embedded in our senses is a type of spatial awareness, a proximity to our bodies. For instance, we can see things in a distance, we can hear things that are very far, we can smell things that are fairly close and we can only feel things that we are directly touching or that are in very close proximity. Sound, like touch is ephemeral, sound is constantly washing over us and our brains must filter out many sounds choosing what specifically to remember (Ahrens et al. 2015). Apart from that, we each form memories of specific events in slightly different ways, which is to say that we each have our own perspective and our own vantage point to any given event. Interestingly, in the collection of multiple data sets of several people's memory of the same event, there is always a micro and macro level of the event itself.

Lancel and Maat have been keeping the larger body of data from each kissing event in their experiment. This makes their project a large reservoir (database) of human intimacy with a trend-setting potential (we may all want to have our own digital kiss one day). Like digital signatures, digitally recorded and preserved gestures of affection stand a great chance of becoming a standard of intimacy – a new form of expressing emotions beyond the limitations of the body. Asked if the sonic is to become a new plane for intimacy and proof of love, the artists respond:

> We believe that art, intimate encounter and public space share a vital interest in facilitating an ongoing personal and shared process of sense-making. Sharing direct and disrupted connections between people involved in intimate touching, datafication and spectators enables authentic intimate experience in public space of merging realities. Important to add, is that instead of validating data, we found that spontaneous data sonification enhances such personal and shared sense-making. There for, spontaneous and unpredictable EEG sound interaction between participants is part of the design. (A few Actors, who indicated that they tried to control the sound through different ways of kissing, in fact acting as if controlling a game, referred to their kiss as fun rather than as intimate.) When discussing the data in dialogue with the host, people's expressions describe 'an enigmatic mirror of their kisses': 'It leaves sense making to ourselves' and 'Only we know what these sounds and traces mean' – appropriating the sonic kiss data, indeed, to plane through intimate encounters. (Lancel and Maat 2019: n.pag.)

CONCLUSION

Sonification is a new frontier in computational aesthetics. While music has been around for thousands of years, the direct sounding of data is a fairly new endeavour with our earliest examples only dating back 100 years. And while sonification lacks the history of music, it presents a compelling new use for an old process. As an extension of music, sonification provides a new contextual framing of an old technology. The thoughts and ideas that surround conventional processes of organizing sound are evolving. *EEG KISS* is one such example of this new path forward in the organization of sound that can provide insight and context to abstract flows of data and information.

As a space for artistic creation, audio-visual representation of data can provide a deep insight into ourselves as a space for poetic exploration. And while this is a time-based medium, it can anchor thoughts and emotions in the minds of those observing the work. The audience is a crucial element of the work and forms a crucial part of the framing of the experience. To be able to hear and see the emotions of the kissers, the viewership of the audience is needed to amplify the emotions of the kissers. If the same kiss were to happen on the street or in a closed room with no one else present, the output would be very different.

The abstraction of the brain data of the kiss in *EEG KISS* creates an intimacy within the work. While we have audience members who are augmenting their bodies with technology, the overall result is in an opening vulnerability and humanization of the kissers. Technology, in its worst cases, can be used as a force of hegemonification as looked at by both Benjamin (1968) and Adorno and Horkheimer (2016). But in its best cases, it can open up new ideas and new ways of understanding our world, including our emotional performance.

References

Adorno, Theodor W. and Horkheimer, Max (2016), *Dialectic of Enlightenment*, London: Verso Books.

Ahrens, Sandra, Jaramillo, Santiago, Yu, Kai, Ghosh, Sanchari, Hwang Ga-Ram, Paik, Raehum, Lai, Cary, He, Miao, Huang, Josh Z. and Li, Bo (2015), 'ErbB4 regulation of a thalamic reticular nucleus circuit for sensory selection', *Nature Neuroscience*, 18:1, pp. 104–11.

Benjamin, Walter (1968), *Illuminations: Essays and Reflections*. New York: Schocken Books.

Lancel, Karen and Hermen, Maat, (2019), interview by author, San Francisco, 20 November.

Lancel, Karen, Hermen, Maat H. and Frances Brazier (2019), '*EEG KISS*: Shared multi-modal, multi brain computer interface experience, in public space', in A. Nijholt (ed.), *Brain Art: Brain-Computer Interfaces for Artistic Expression*, Verlag: Springer Verlag, pp. 207–28.

Pettman, Dominic (2017), *Sonic Intimacy. Voices, Species, Technics (or How to Listen to the World*, Stanford: Stanford University Press, pp. 207–28.

Wlodarski, Rafael and Dunbar, Robin I. M. (2013), 'Examining the possible functions of kissing in romantic relationships', *Archives of Sexual Behavior*, 42:8, pp. 1415–23.

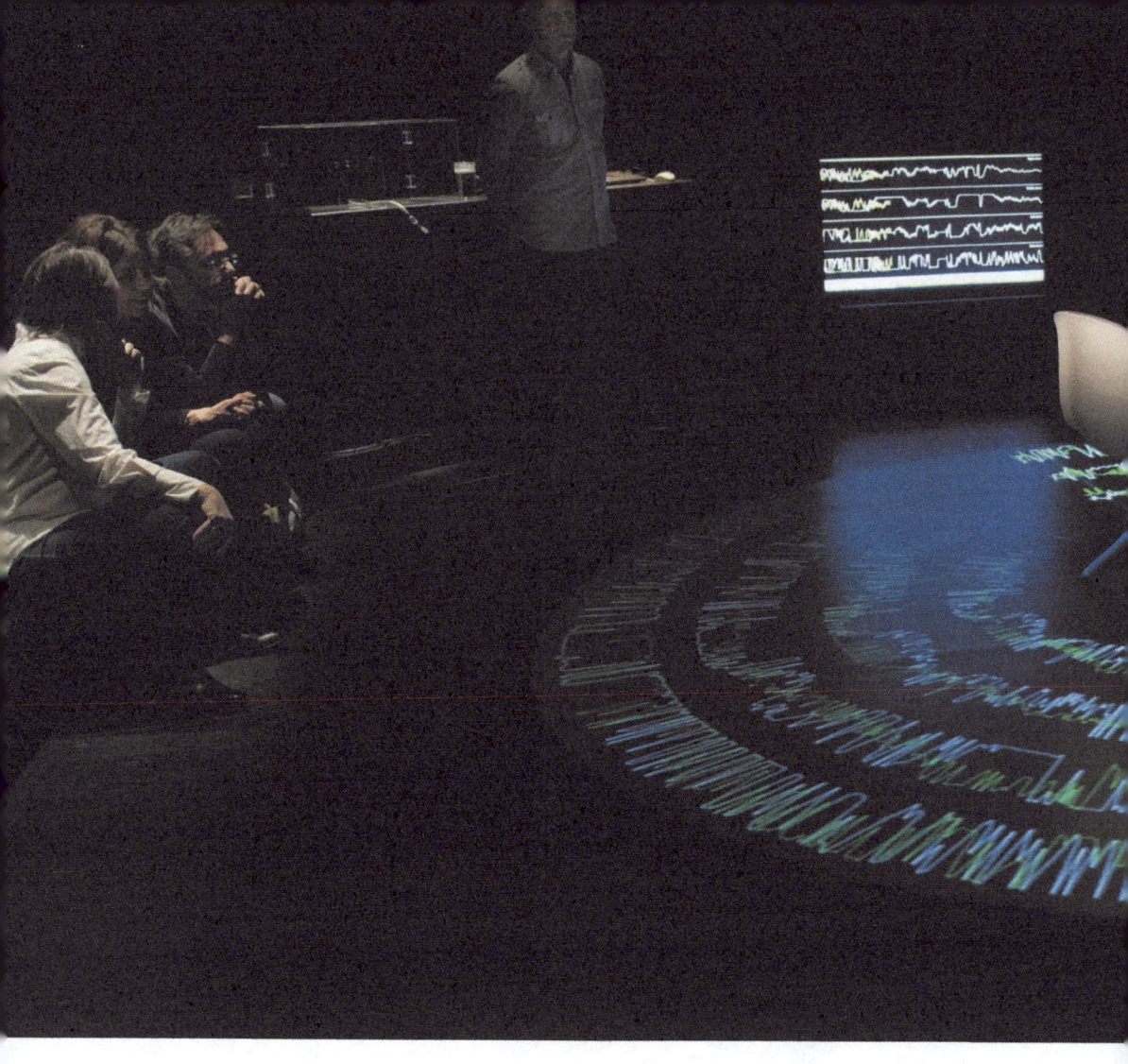

FIGURE 9.3: Lancel/Maat, *Digital Synaesthetic, E.E.G. Kiss Portrait*, 2016. Audiovisual performative installation, frascati Theaters. Courtesy of the artists (Photo by Anna van Kooij), © Lancel/Maat.

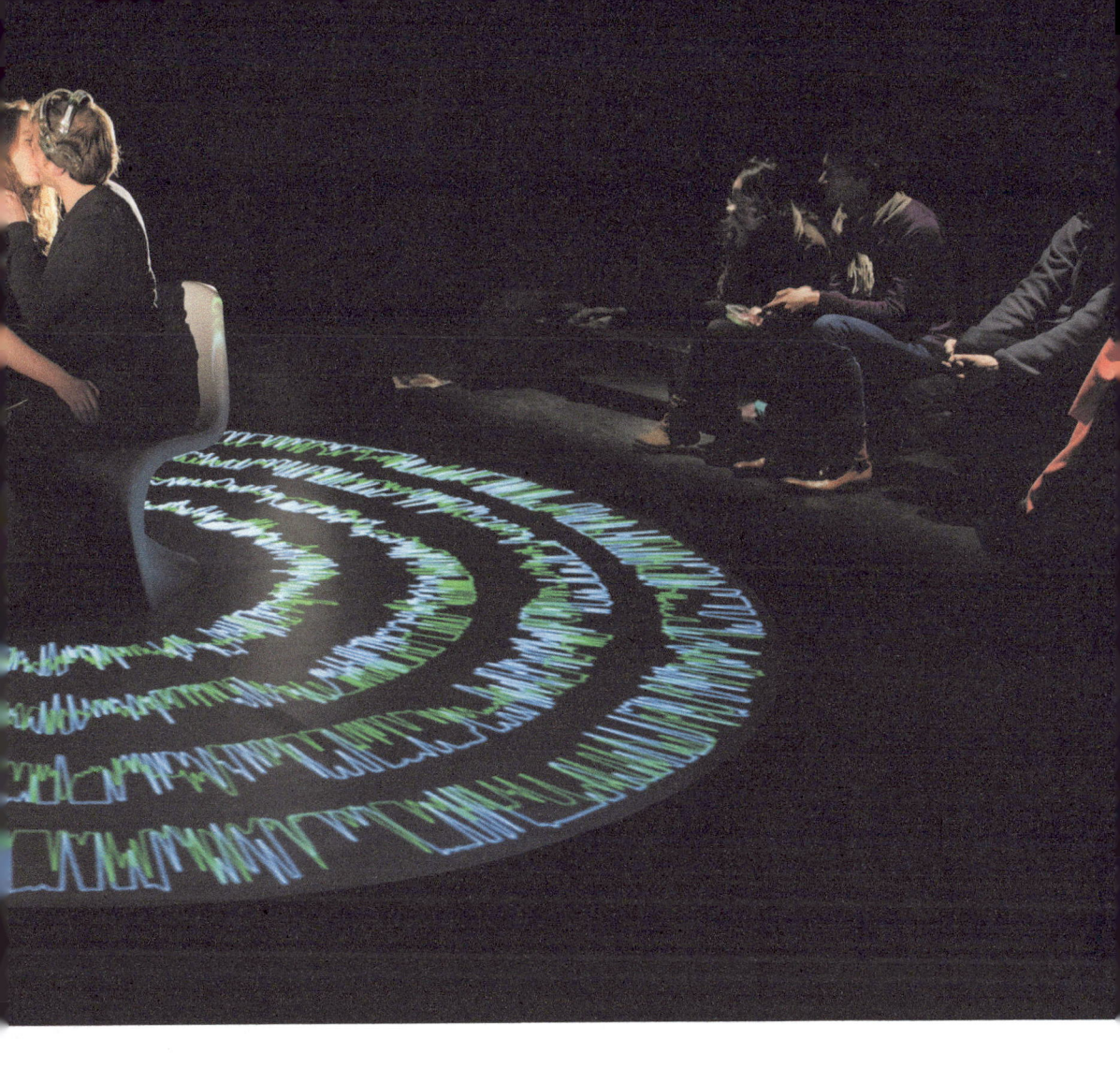

10
Interfaces of Emotional Surveillance

Almost all online messaging systems today – be it IOS Messages, Facebook Messenger, WhatsApp, We Chat, Line or others – offer tracking services to notify the users on the interaction status on both ends of a 'chat room'. This innocent feature has quickly turned into a surveillance measure that would allow the chatting parties to track their mutual engagement in the conversation process and, based on that, assess the dynamics of the relationship outside of the conversation. The feature has been of common use by romantic couples (daters, spouses, lovers) and is now an integral part of the romantic process, overexploited for predicting (and forestalling) stages of a relationship. Like geolocation services that help people 'map' their whereabouts (in a literal and metaphorical sense), message tracking serves to map the coordinates of a relationship and its status. The questions these services inspire are not: 'where are you' and 'when were you last active', but 'where are you with me in the relationship' and 'how active are you there with me'.

The role of messages tracking for romance has been discussed in psychological and sociological literature in terms of the new signifiers of involvement and care. Research reports a lot of emotional drama over location services and timestamps. For instance, unreturned text messages stamped as 'received' or 'seen' evoke suspicion, jealousy or mistrust – especially when the partner who does not respond is still visible online. Suspicions of infidelity and waning commitment are common in couples who monitor each other on GPS apps. New forms of anxiety and compulsive obsession emerge from controlling a partner's moves and online activeness. Online surveillance is now an element of dating/flirting practices, anchored in our deepest and darkest insecurities about love. It brings out culturally induced fears that we re-enact with the media cultivating a new idea of coupledom.

Tom Galle and John Yuyi brood over digital surveillance in *Face Messenger* – a close-up of an eye, captioned 'seen 2:28', the caption inked under the eye in a font typical of the timestamp. Next to the timestamp, there is a tick symbol to mark the reception of the message. The gaze frozen in some resigned exhaustion informs about three defining states of an online 'watch': alertness (vigilance over the device), control (the monitoring of the communication flow) and antic-ipation (a suspended excitement of waiting for a response). While this intense gaze awakens imageries of surveillance and power-games, it also brings up the mysterious tradition of 'Lover Eyes', which started in the eighteenth-century Britain when Prince George of Wales sent Anne Marie Fitzherbert a miniature painting of his own eye. This gesture was believed to symbolize affection and care. The 'Lover Eye' was mostly in a form of jewellery – rings, necklaces, brooches – that lover's exchanged as a sign of commitment and a reminder of an unrelent-ing desire for the presence of another. Today, the mobile phones like 'Lovers Eyes' transmit our digitalized gazes across the screens and aether, trying to keep the connection always on and in a permanent state of anticipation. Romantic interactions run a lot on 'uncertainty reward' that flushes our brains with dopamine. The new media enhance the effect keeping the subjects in a permanent state of excitement, exhausting in its constant availability and unescapable tether.

Face Messenger

TOM GALLE AND JOHN YUYI

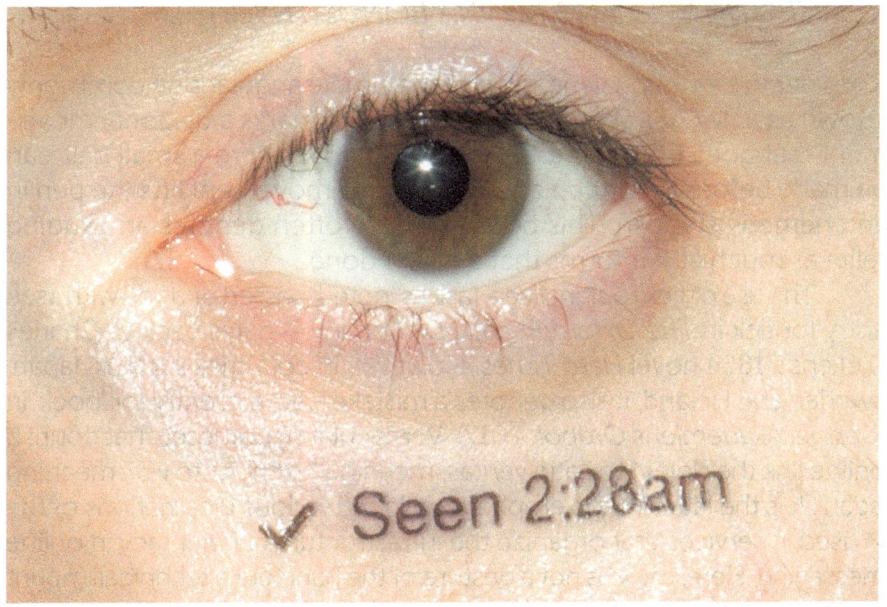

FIGURE 10.1: Tom Galle and John
Yuyi, *Face Messenger*, 2016. Acrylic
Print. Courtesy of the artists.

The Uncertainties and Anxieties of Timestamps and Read Receipts in Online Messaging

KRISTIN VEEL AND NANNA BONDE THYLSTRUP

Consider the emotional meaning of the simple angular sign commonly known as a tick (✔). When drawn by hand, it marks a diagonal movement that goes from the top left downwards, where a small halt can be made before a swifter diagonal upwards movement lifts the pen in an energetic swoosh. This bodily gesture often denotes an exuding relief as much as it informs that a task is done.

The use of tick to signalize a task as 'noted, passed or done with' (see entry for tick in *The Oxford English Dictionary*) – dates back to Charles Dickens's 1854 novel *Hard Times*. However, in countries such as Japan, Sweden and Finland, it also denotes a mistake (see, e.g. entry for 'bock' in *Svenska Akademiens Ordbok* 2017). Vernacular hypotheses that flourish online link the sign ✔ to Latin *veritas*, meaning 'truth', or to *vidi*, meaning 'seen'. It is the latter that is emphasized in the various incarnations of the ✔ used in services that organize the infrastructures of intimacy in online messaging. Here, the ✔ is not a gesture of the hand but a technical imprint often coupled with the word 'seen' and a timestamp.

However, a very different notion of control, monitoring and directionality is at work in the timestamp use of a tick. While a timestamp can still indicate the completion of a task by informing the sender that their message has been sent to or received by the addressee, the timestamp also immediately marks a state of anticipation at both ends. The sender anticipates a response, and the receiver broods over the response – how and when to reply or if to reply at all. The use of the timestamp, and its related forms of agency, signifies an anticipation (pointing towards future horizons) as much as it signifies the completion (endpoint) of something. Although the timestamp gives us an insight into the process of information exchange, it also creates space for uncertainty and suspense that can induce a sense of anxiety in the participants of an exchange. Those emotions intensify with other information markers such as activity statuses (online, offline, idle, active) or information on users' whereabouts. The latter is available through geolocation data, mobilized, for instance, in dating apps (Veel and Thylstrup 2018) and other forms of geosocial networking.

This chapter analyses the 'emotional politics' of timestamps and location tags in online messaging. It moves from discussing the dynamics of anticipation with its uncertainties and anxieties to exploring the ways in which we are all *captured* by the system (to use Philip Agre's terminology). It ends with the examination of fundamental inconveniences imposed by the status of 'active now', including the technical properties of the timestamp, which can obscure the sense of agency in the condition of constant visibility.

We focus here on the management of relationships and on how media functions, such as timestamps and activity statuses, inscribe themselves in the execution and maintenance of flirtation. Like offline hookup, online dating is riddled with uncertainty and interpretative surplus even in the most uncommitted and superficial engagements. This is because the two interlocutors who do not know each other very well communicate within a very limited range of data, cues and information. This lack of knowledge can be exhilarating, agonizing and antagonizing – all at the same time. And because the unknown is key in flirtation, the unknown may even be part of the drive. As psychoanalyst Adam Phillips (1993: 54) notes, when 'flirt[ing] with possibilities, we are both the hunter and the hunted'. Romantic relationships are a particularly good ground for a close reading of the affective dynamics of online messaging, especially in terms of control (Donghee, Carr and Hayes 2016; Hayes, Wesselmann and Carr 2018). Yet, the same considerations may also apply to other areas of online communication insofar as online flirting mimics the general logic of online surveillance and tracking by often inducing more uncertainty and tension, also on a societal level (see, for instance, Natasha Dow Schüll's work on the affective dynamics of the uncertainties of gaming and gambling: [2014]).

SEEING AND BEING SEEN

In his recent article for *Dazed Digital*, journalist Jack Palfrey notes,

> There's the gut-wrenching and humiliating emotional sucker-punch of knowing someone's seen your message but not replied straight away. We're left questioning what we've done wrong or, probably more likely, furious at the receiver for having *the audacity* to not drop what they're doing and get back immediately. We construct entire, baseless narratives around that 'seen' notification. Sometimes, we use them passive aggressively, ruthlessly informing people we're unhappy with radio silence. (2017: n.pag.)

Palfrey is not alone in his negative assessment of the 'read receipt'; Firman (2017) believes that '"seen" notifications make social media even more addictive'. It is highly unlikely that the inventors of the timestamp and the read receipt caption intended any anxious responses to the otherwise subtle graphical user (GU) interface.

Whatever motivation behind their design, the origins of timestamps are difficult to track and lack a proper historical treatment. In this respect, the timestamp shares the fate of other unobtrusive media 'small forms' (Vogl n.d.), such as the thumbnail image (Thylstrup and Teilmann-Lock 2017). One way to trace those forms' emergence within social digital environments, however, is to attend to patents that lay property claim to their origin (Hemmungs Wirtén 2019). The patent description for timestamp offers an insight into a data-dissemination procedure and information on how that small form of social media serves to control, reflect and channel processes of communication. This makes it clear how the timestamp has always worked to actively transmission experiences, as well as how it has sought to steer and saturate curiosity.

The recent patent war between Blackberry, Facebook and Snapchat has proven that timestamps are both ubiquitous features in social media messaging systems and also a subject to contestation over origins (Macias Jr. 2018). Blackberry claimed that Snap's messaging systems had infringed upon its patent for 'handheld electronic device and associated method providing time data in a messaging environment' (see Klassen et al. 2012). Although timestamps appear in the patents of many other digital communication systems, Blackberry's patent seems like a good place to start tracing the timestamp's logic. Here, the inventors note that how the recent invention of 'instant messaging' has created a new demand for displaying communication threads to both the sender and the receiver. In this new messaging environment, they observe that a need has thus arisen to inform those engaged in a conversation on the interruptions in the flow of the conversation. Interestingly, the Blackberry's patent situates the use of this function in a romantic context.

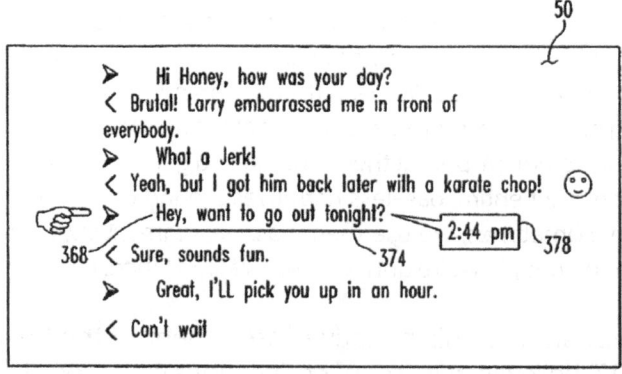

FIGURE 10.2: Illustration from Klassen 2012 et al. Patent No.: US8,301,713B2.

Yet, while the incentive for adding timestamps to online messaging services seems to have emerged from the need of counteracting miscommunication, it quickly became clear that the level of precision offered by this service creates more uncertainty than it alleviates. The problem with the message '✓ Seen' (and its accompanying time tag, e.g. '2:28 a.m'.) is its own systemic imprecision. 'Read receipt' is no guarantee to the sender that the receiver has actually read their message. It only tells them that the message has been automatically registered by the system, and that the system generated the time tag in response to the activation of the screen. The rest is a total mystery, and the actual meaning of 'Seen' and its whereabouts remains secret until the response arrives – if ever. By then, you never know whether the message was actually read (or opened by accident), was the content anticipated and welcome with enthusiasm or with indifference. Those uncertainties are of course endemic to any form of message writing. All fiction and cinema that feature communication technics and technologies – from letters to the telegraph to online messaging services – use messaging as a way to create narrative suspense in a storyline. However, communication infrastructures in digital media offer different dynamics from traditional epistolary practices and analogue messaging. Indeed, the way in which the timestamp tries to alleviate uncertainty by increasing the level of detailed information can be seen as characteristic of our present cultural moment. Yet, as we know from the hype of big data, more information does not necessarily increase our sense of control. On the contrary, it allows for new epistemological cracks that gesture towards new unknowns, beyond our control (Agostinho et al. 2019). WhatsApp, for instance, has introduced a two-stage notification with one and two ticks, where ✓means 'delivered' and ✓✓ means 'seen'. Yet, as much as this system of one or two ticks specify the stage of the delivery process, it also foregrounds the complexity of the interrelations between messaging services and people, by pointing to the distinction between human and non-human ways of seeing a message (Zylinska 2017). Messenger, Signal, iMessage and the various messaging services of online dating apps and platforms use different icons for distinguishing between delivery and 'seen'. The logic, however, is the same and so are the frustrations about the new uncertainties introduced by the read receipt.

Yet, there are also fundamental similarities between the digital message-delivery systems and their historical predecessors. Many of them help us understand what is specific about their online iterations. Letters, for instance, have always needed a trusted carrier, leaving the sender and the receiver in a state of prolonged anticipation. An organized postal service (which can be dated back to ancient Egypt) would establish a system of delivery updates, including an approximate delivery time for a given distance or geographical location, all of which might help alleviate the agony of waiting for a response. Later, new technological inventions such as telegraphs and telephones pushed messaging towards both immediacy and receipts of delivery. In a US postal context, the United States Postal Service introduced Registered Mail in 1855, which essentially would give a sender a receipt of delivery if they requested one. In 1955, a century later, the USPS rolled out Certified Mail, which offered the sender certainty of delivery by giving them a literal receipt. And more recently, new delivery companies such as UPS and DHL offer senders the possibility of tracking the entire route progress of a package or letter.

New forms of anxieties have emerged from these possibilities, engendering forms of anticipation in which anxiety reverberates the rhythms of instantaneity. Contemporary online messaging combines notifications, communication and feelings in intricate ways to grant the possibility of an immediate response as well as micro-tracking (e.g. the timestamp), enabling the access to ever more detailed information about the message's route. Delivery time as a locus for anticipation is thus supposedly annihilated in digital space. In this optimized realm of communication, uncertainty attaches itself not so much to whether a message has made it to the receiver but rather to the temporal gap between knowing that a message has been 'delivered: and knowing that it has been "seen"'. Latency is thus more or less relinquished to the human factor (rather than a technological feature of a messaging software of a device); any delay depends on interlocutors and is grounded in the nature of their relationship and their knowledge (if any) of communication styles and life routines.

How might we understand the management effects of the timestamp as a cultural problem of visibility and exposure? In *Being and Nothingness: A Phenomenological Essay on Ontology* from 1943, in which Jean Paul-Sartre explores consciousness and perception as part of the development of his existentialist philosophy, he (1966: 347) describes in a seminal scene how he is looking through a keyhole wholly engrossed in what he sees and not conscious of himself looking. However, when footsteps are suddenly heard in the hall, he becomes aware of being seen: 'I see myself because somebody sees me' (Sartre 1966: 350). Sartre's point here is that when we become an object of another's look, we lose autonomy as we find ourselves seen and interpreted by the other. In online messaging, although the agency resides with the person that is responding, the emotions and affects related to having been seen reading the message and the *anticipation* of a response are with both the receiver and the sender. This opens space to power games and may indeed result in a power tug-of-war, contradicting the pleasurable aspects of online interactions (and their real-life follow-ups), which thus prompts us to respond quickly in order to escape the situation and regain control.

People's reactions to read receipts reflect serious ambivalences and anxieties about the visibilities of message statuses, not least in dating platforms. When Tinder first launched read receipts, notifications for that feature were a default part of the service. After complaints from the users (mostly women) and reports on the function's negative psychological effect, the receipts were temporarily removed (Jamie 2019) to return as a paid service in 2010. The receipts can now be purchased by groups of five, ten or twenty, and activated for a particular conversation. It, however, requires an activation on both sides; in other words, to use it effectively, the person you are messaging needs to activate this function in their settings menu. If they turn this function off, it will not be possible for others to see the reception status of their message, even if they have bought the service. The trajectory of the 'read receipt' on Tinder speaks volumes about the power dynamics related to knowing whether your interlocutor has seen your message and the importance of knowing that they are aware of having been seen. Conversely, it can be argued that there is a willing embrace of vulnerability in allowing others to follow your reception status (which can also be interpreted as a resourcefulness that enables some to rise above communication anxieties alongside a logic of 'nothing to hide and nothing to fear'). One way or another, the reception status itself becomes a commodity that potentially trades the control in the messaging process for one's emotional well-being and vice versa.

THE TIMESTAMP AND GEOLOCATIONS: CAPTURING DATA AND EXPOSING SUBJECTS

The timestamp is considered a small form (Veel 2021) inherently oriented towards novelty. Indeed, as Benjamin Grosser (2014) observes, the timestamp itself is a 'language [that] represents an engineering of ideology, an algorithmic realization of a preference for the new'. Hence, we frame timestamps as essential social instruments of calibration that feed into larger systemic processes of optimization and efficiency. In these processes, the confirmation that a message has been delivered and read comes to represent a functionality of technologies preoccupied with ensuring ever-smoother communicational transactions.

Speaking of the usefulness of the timestamp in the display of electronic communication, inventors Flynn and Foran point out that

> [b]usiness people and others need to verify that an important transaction once sent has been received by the intended recipient. The main obstacle to widespread commercial use of electronic communications, such as for example email and email attachment, is the lack of the ability to verify that the email and/or attachment was received by the intended recipient. Email must be sent on unsecured pathways, pathways where the email can be mis-directed, lost, and/or altered. It is highly desirable to the sender to be able to verify that the intended recipient has received an important email. It is also desirable to the sender to know that the intended information in electronic message was received as written or sent. (2003: 1–2)

Timestamps and read receipts are also often accompanied by a location tag if the conversation partner(s)' location option is enabled in their communication device. The integration of such optimization technologies embeds the individual into larger geo-temporal architectures. As Sarah Sharma notes, the new media-bound reality calls for a constant recalibration of 'accounts for the multiple ways in which individuals and social groups synchronize their body clocks, their senses of the future or the present, to an exterior relation – be it another person, pace, technology, chronometer, institution, or ideology' (2014: 18). In this system, the geo- and timestamps are both technologies of communication and transit spaces where messages, persons, data and capital circulate. As such, time- and geostamps belong to what information theorist Philip Agre (1994) calls 'capture systems'.

To Agre's views, capture systems describe the logic of tracking systems such as UPS barcodes, the Global Positioning System, RFID and others. In his view, these have two primary functions. The first function 'refers to a computer system's (figurative) act of acquiring certain data as input, whether from a human operator or from an electronic or electromechanical device' (Agre 1994: 107). The second function is 'a representation scheme's ability to fully, accurately, or "cleanly" express particular semantic notions or distinctions, without reference to the actual taking in of data' (Agre 1994: 107). These functions together form what Agre calls 'grammars of action', i.e. symbolic expressions of computable procedures. Wendy Chun (2019) has foregrounded Philip Agre's germinal work on capture systems as a model for understanding what goes on in the temporal architectures of instant messaging (Chun 2016), and it therefore may help us to unpack the logic of the timestamp in dating apps too.

As Armin Beverungen, Timon Beyes and Lisa Conrad (2019) note, Agre's examples are taken from the world of organization: accounting systems, telemarketing, user interfaces or enterprise integration systems. Yet, Agre also emphasizes while such systems are often deployed to track objects in large organizations, one should not seek to make a distinction between systems that track humans and those that track objects: both rely on the same logic of capture. As he notes, 'a system that tracks people by means of an identification card, for example, is really tracking the cards' (Agre 1994: 742). It is therefore no coincidence that Agre correlates the logic of capture with the deeper forces of marketization and democratization (Agre 1994; see also Chun 2016). Drawing from Ronald Coase's theory of the firm, Agre argues that once a grammar of action has been imposed on an activity, discrete units and individual episodes of that activity are more readily identified, verified, counted, measured, compared, represented, rearranged, contracted for and evaluated in terms of economic efficiency. In a sense, then, not only do capture systems capture movements, but through this capture, they also enable a certain empowerment based on the marketization of all transactions. Importantly for our understanding of dating apps, capture systems can only compute with what they capture – so the less a dating app captures, the less functionality it offers to its user (Agre 1994: 749). The timestamp and geostamp are important elements in such capture moments, offering the system and its user knowledge about others. At a more fundamental level, however, the empowerment offered to the user by means of information obtained through tracking does not match the level of empowerment offered to the company doing the tracking. The latter allows the company to harvest ever more granular information about dating habits but also about personal networks, consumption patterns and intimate details.

This imbalance in access to information has led to a prevalent discourse about surveillance in dating apps and a sense that datable subjects in those apps are subject to a double surveillance both from other users and from the overall system of capture. Yet, if we stick with Agre's argument, capture systems cannot be likened to surveillance systems in a traditional sense. To support his argument, Agre points out that while surveillance systems rely on visual/territorial metaphors and are centralized systems linked to state power, capture-based systems, in contrast, rely on linguistic metaphors and constant movement, are distributed systems and are linked to private corporations. As Wendy

Chun observes, capture and surveillance systems also differ in terms of their base units of analysis and encourage different forms of practice:

> In a capture system, the base unit is an action, change of state, rather than the entire person. It enables a finer grid by presuming and enabling mobility: in order for something to be captured, it must be in motion. There must be a change of state. Things must be updated in order to register. Because of this constant activity, people who engage in heavily captured activity have a certain freedom, namely free creation within a system of rules. (Chun 2016: 60)

While Agre's clear distinction between the different logics at play in surveillance and capture rests on a too sharp epistemic division between state and corporate power, his system nevertheless helps us to understand the different logics at play in the timestamp-dating apps: on the one hand, they become modes of surveillance (as 'seen' messages), and on the other hand, they also operate as temporal mechanisms of capture (as temporal architectures of the new). Most notably, then, Agre's distinction helps us to see the timestamp as an instrument of intimate surveillance as well as a mechanism of optimization, where what is being 'seen' is both a subject and a change of state. The more we swipe, tap, interact, move and search, the better the system can register us, offering the freedom to create more possibilities but crucially also enclosing us within ever more complex systems of rules (Møller and Petersen 2017).

THE UNCERTAINTY OF MICROMANAGEMENT: AGENCY IN A BROAD PRESENCE

For the user, the timestamp gives the impression of accuracy. It freezes the moment when a message was sent, delivered or seen, in a way similar to the postmark, which is still often used today as a proof that a letter has been dispatched, even if the delivery was delayed. The timestamp is also used as authentication in a number of systems – for instance, in blockchain technologies where timestamps mark the chronology of the nodes on the blockchain and thus serve as proof of when and what happened in the chain. In online messaging services (especially those focused on dating), it may take on similar functions (as evidenced by the re-establishment of the read receipt on Tinder in 2019). Yet, as we look closer at time measurement systems, questions emerge about their skills for precision and thus their authority as authentications. For instance, if you are on a plane and put your phone into a flight mode, it will not be able to send a read receipt, even if you see the message on your screen. Depending on the configuration, your phone will try repeatedly to send the read receipt and will eventually do so once you land and are able to connect to the network again, or alternatively, it will simply give up, and your message will keep appearing as unread even though you have in fact seen it. What is at stake here is the translation process between system time and Coordinated Universal Time (UTC). Most devices operate with system time, which is a way of calculating a set of intervals from an arbitrary starting date. For instance, the Unix time system calculates the number of seconds elapsed since 00:00:00 UTC on 1 January 1970, with each day containing 86,400 seconds. Other systems that start from different dates and calculate with other intervals abound. However, in order to become useful for the human user, these can be converted into UTC by way of subroutines that also adjust for time zones, daylight saving time and what are known as 'leap seconds'.[1] The timestamp we see and interact with is thus a different one from the system timestamp, but the system timestamp is what is at the heart of the logic of the timestamp, and also what mobile forensic work seeks to extract (Mahalik et al. 2018).

These complex translation processes are constantly going on behind the simplicity of the timestamp you see in your online messaging service, but it is rarely critical in the everyday management of online exchanges. However, it speaks to the ambivalence of the accuracy of time measurement, which becomes ever more acute as the time span between interactions becomes shorter. Online messaging services – whether it is iMessage, WhatsApp, Signal or the chat functions connected to dating apps – are often used essentially as instant messaging for real-time conversations that take place in an extended now, where micromanagement is of key importance. This is emphasized by the way we speak online. Sociolinguistic approaches to social media narratives argue that the

> real-time narration found most prominently in social-network sites like Facebook and Twitter favours present-tense or non-finite verb forms, creating an ongoing sense of an ever-present 'now' that bridges the asynchronous gap between the time of narrative production and narrative reception. (Page 2013: 41)

The same can be argued for online messaging, which bears the linguistic features of a face-to-face conversation in real time. There is a linguistic anticipation of an immediate answer, which only increases the impetus to answer immediately. This turns the timestamp into an awkward conversational silence that can only be ended by a response – indeed any response, even a meaningless emoji. The real-time experience in this way creates an extended now that is emphasized by your ability to see whether the person with whom you are communicating is 'online' or 'active now' – terminology that contributes to the sensation of a broad present where interactions play out (Cox and Lund 2016; Ernst 2016; Gumbrecht 2014; Osborne 2014; Steiner and Veel 2020). The ability to see whether the other person is texting, through the icon of moving dots featured in many online messaging services, adds to this sense of real time.

The question remains as to the degree of agency found in this situation where you are writing in the knowledge that you are being seen by both your interlocutor and the capture system. As media scholar Tara McPherson wrote of online 'liveness' as early as 2006:

> This liveness foregrounds volition and mobility, creating a liveness on demand. Thus, unlike television, which parades its presence before us, the Web structures a *sense of causality* in relation to liveness, a liveness which we navigate and move through, often structuring a feeling that our own desire drives the movement. The Web is about presence but an unstable presence: it's in process, in motion. (McPherson 2006: 202, original emphasis)

The sensation of real time creates a sense of urgency as well as agency. This is exactly what encourages us to be constantly online and what propels the process of immediate responding. After all, not responding to someone on a messaging service (often colloquially referred to as 'ghosting', see: Safronova 2015) might be revealed as such if you are actively present on other platforms. Significantly, however, because captured subjects enjoy a certain freedom – namely, free creation within a system of rules – they can also optimize their actions by gaming the system (Chun 2016: 60). An article in *Vox* advises on how to game the system noting that 'the first step is to understand that Tinder is sorting its users with a fairly simple algorithm that can't consider very many factors beyond appearance and location'. It then moves into detailed ways of acquiring more likes and optimizing one's standing in the system through representation and activity before concluding:

> Once you sift through those and winnow out the duds, you should be left with a few solid options. If not, go back to swiping but stop again at nine. Nine is the magic number! Do not forget about this! You will drive yourself batty if you, like a friend of mine who will go unnamed, allow yourself to rack up 622 Tinder matches. To sum up: Don't over-swipe (only swipe if you're really interested), don't keep going once you have a reasonable

number of options to start messaging, and don't worry too much about your 'desirability' rating other than by doing the best you can to have a full, informative profile with lots of clear photos. Don't count too much on Super Likes, because they're mostly a money making endeavour. (Tiffany 2019: n.pag.)

The author of the article also notes other ways of gaming the system, for instance, by lurking on old secondary-school acquaintances and speculating on their datable potential today:

I don't think you can get in trouble for one of my favorite pastimes, which is lightly tricking my Tinder location to figure out which boys from my high school would date me now. But maybe! (Quick tip: If you visit your hometown, don't do any swiping while you're there, but log in when you're back to your normal location – whoever right-swiped you during your visit should show up. Left-swipers or non-swipers won't because the app's no longer pulling from that location.) (Tiffany 2019: n.pag.)

Like in the case of the 'read' receipt when your phone is in a flight mode, here we see ways in which a capture system can be gamed. Articles advising users how to navigate and optimize their chances often read like manuals on how to get the most out of the system as an individual user, but they also constitute prime examples of how the more general state of uncertainty induced by the 'dating game' generates a series of micromanagement procedures. Those procedures simultaneously contribute to an increase in uncertainty *and* offer avenues for playfulness, which may ultimately place you differently not only in your own individual dating game but also in the way the network monitoring your actions, analyses and responds to those actions.

CONCLUSION

The timestamp is part of a tendency in contemporary capture and surveillance societies to capture in ever finer details when and where movements occur and these trackings invariably become part of ongoing power negotiations. It may be argued that the timestamp adds a sense of accuracy and introduces a negotiation of agency and mastery, which acquires its full significance when a message is not immediately read or responded to. Significantly, in most messaging services, the timestamp is only visible for the last message sent, and it disappears from the display as soon as a response is registered. In that sense, the

✓ can indeed, as we have discussed in this article, be read as signify-ing a 'job done': the ✓ is a gesture that belongs not to the receiver of the message, but to the person who sent the message, and who can now consider the ball in the other person's court until a new response arrives. However, as we have seen, the ✓ in the timestamp takes on a new temporal and spatial meaning – fraught with uncertainty and urgency, rather than designating information that can be norma-tively valued as right or wrong (depending on the cultural context in which the sign appears). Building on Doane's (1990) and Feuer's (1983) canonical work on the temporality of television, Chun pinpoints the difference between 'using' online media – which operates through a perpetual series of *crises* that demand a real-time response and 'watching' a *catastrophe* live on TV:

> Crisis moments that demand real time response make new media valuable and empowering by tying certain information to a deci-sion, personal or political (in this sense, new media also personal-ises crises). Crises mark the difference between 'using' and other modes of media spectatorship/viewing, in particular 'watching' television, which has been theorised in terms of liveness and catastrophe. Comprehending the difference between new media crises and televisual catastrophes is central to understanding the promise and threat of new media. (Chun 2011: 95)

Being online fosters the sense of a series of ever-present crises that we need to react and respond to. As Grosser (2014: n.pag.) notes:

> By relentlessly reminding us of the age of a post relative to the present, timestamps create a false sense of urgency. [...] But these constant enumerations of age present the news feed as a running conversation that you can't miss – if you leave for even a second, something important might pass you by.

Such a state fosters the need to constantly be online and to have systems that aid us in tracking and thus manipulating our own and others' activities in the digital sphere. Online messaging in its most stressful iteration can be seen as a perpetual series of minor crises that need to be dealt with and responded to, generating a constant stream of activity that in turn feeds into ever more comprehensive systems of capture. Yet, as we hope we have shown, the numerous levels of uncertainty with which these capture systems are fraught also potentially open up small avenues for gaming the system, even while we are caught up in it as *seen* and – if not by the other, then by the system.

References

Agostinho, Daniela, D'Ignazio, Catherine, Ring, Annie, Thylstrup, Nanna and Veel, Kristin (2017), 'Uncertain archives: Approaching the unknowns, errors and vulnerabilities of big data through cultural theories of the archive', *Surveillance and Society*, 17:3&4, pp. 422–41.

Agre, Philip E. (1994), 'Surveillance and capture: Two models of privacy', *Information Society*, 10:2, pp. 101–27.

Amoore, Louise and Piotukh, Volha (2015), 'Life beyond big data: Governing with little analytics', *Economy and Society*, 44:3, pp. 341–66.

Beverungen, Armin, Beyes, Timon and Conrad, Lisa (2019), 'The organizational powers of (digital) media', *Organization*, 26:5, pp. 621–35.

'bock, sbst'. (1917) *Svenska Akademiens Ordbok*, Stockholm: Svenska Akademien, http://www.saob.se/artikel/?unik=B_3521-0198.H986. Accessed 25 November 2019.

Chun, Wendy Hui Kyong (2011), 'Crisis, crisis, crisis, or sovereignty and networks', *Theory, Culture, and Society*, 28:6, pp. 91–112.

Chun, Wendy Hui Kyong (2016), *Updating to Remain the Same: Habitual New Media*, Cambridge, MA: MIT Press.

Chun, Wendy Hui Kyong (2019), 'Queerying homophily', in C. Apprich, W. H. K. Chun, F. Cramer and H. Steyerl (eds), *Pattern Discrimination*, Minneapolis: University of Minnesota Press, pp. 59–99.

Cox, Geoff and Lund, Jacob (2016), *The Contemporary Condition: Introductory Thoughts on Contemporaneity and Contemporary Art*, Berlin: Sternberg Press.

Doane, Mary Ann (1990), 'Information, crisis, catastrophe', in P. Mellencamp (ed.), *Logics of Television: Essays in Cultural Criticism*, Bloomington: Indiana University Press, pp. 222–39.

Ernst, Wolfgang (2016), *Chronopoetics: The Temporal Being and Operativity of Technological Media*, Lanham: Rowman and Littlefield.

Feuer, Jane (1983), 'The concept of live television: Ontology as ideology', in E. A. Kaplan (ed.), *Regarding Television: Critical Approaches*, Washington, DC: University Press of America, pp. 12–22.

Firman, Therese (2017), 'Why Facebook's "seen" notifications make social media even *more* addictive', *Well and Good*, 10 November, https://www.wellandgood.com/good-advice/read-receipts-social-media-addictive/. Accessed 1 March 2020.

Flynn, Francis H. and Foran, Jeffrey (2003), *US 6,618,747 B1*, https://patentimages.storage.googleapis.com/00/f3/76/eaf4ed7861ceba/US6618747.pdf. Accessed 1 December 2019.

Grosser, Benjamin (2014), 'What do metrics want? How quantification prescribes social interaction on Facebook', *Computational Culture*, 4:9, http://computationalculture.net/what-do-metrics-want/. Accessed 29 November 2019.

Gumbrecht, Hans Ulrich (2014), *Our Broad Present: Time and Contemporary Culture*, New York: Columbia University Press.

Hayes, Rebecca A., Wesselmann, Eric D. and Carr, Caleb T. (2018), 'When nobody "likes" you: Perceived ostracism through paralinguistic digital affordances within social media', *Social Media + Society*, https://doi.org/10.1177/2056305118800309. Accessed 1 March 2020.

Hemmungs Wirtén, E. (2019), 'How patents became documents, or dreaming of technoscientific order, 1895–1937',

Journal of Documentation, 75:3,
pp. 577–92.

Jamie (2019), 'How to tell if someone read your message in Tinder', *Techjunkie*, 8 November, https://www.techjunkie.com/tell-someone-read-message-tinder/. Accessed 25 November 2019.

Macias Jr., Martin (2018), 'Blackberry patent war mostly survives attack by Facebook, snap', *Courthouse News Service*, 23 August, https://www.courthousenews.com/blackberry-patent-war-mostly-survives-attack-by-facebook-snap/. Accessed 1 March 2020.

Mahalik, Heather, Bommisetty, Satish, Skulkin, Oleg and Tamma, Rohit (eds) (2018), *Practical Mobile Forensics: A Hands-on Guide to Mastering Mobile Forensics*, Birmingham: Packt Publishing.

McPherson, Tara (2006), 'Reload: Liveness, mobility, and the web', in W. H. K. Chun and T. Keenan (eds), *New Media, Old Media: A History and Theory Reader*, London: Routledge, pp. 199–209.

Møller, Kristian and Petersen, Michael Nebeling (2017), 'Bleeding boundaries: Domesticating gay hook-up apps', in R. Andreassen, K. Harrison, M. N. Petersen and T. Raun (eds), *Mediated Intimacies: Connectivities, Relationalities, Proximities*, London: Routledge, pp. 208–24.

Page, Ruth (2013), 'Seriality and storytelling in social media', *Storyworlds*, 5, pp. 31–54.

Palfrey, Jack (2017), 'How "seen" messages on Facebook mess with your mental health', *Dazed Digital*, 9 November, https://www.dazeddigital.com/science-tech/article/38010/1/how-seen-messages-on-facebook-mess-with-your-mental-health. Accessed 1 March 2020.

Phillips, Adam (1993), *On Kissing, Tickling, and Being Bored: Psychoanalytic Essays on the Unexamined Life*, Cambridge: Harvard University Press.

Osborne, Peter (2014), 'The postconceptual condition: Or, the cultural logic of high capitalism today', *Radical Philosophy*, 184, pp. 18–27.

Rosamond, Emily (2018), 'To sort, to match and to share: Addressivity in online dating platforms', *Journal of Aesthetics and Culture*, 10:3, pp. 33–42.

Safronova, Valeriya (2015), 'Exes explain ghosting, the ultimate silent treatment', *New York Times*, 26 June, https://www.nytimes.com/2015/06/26/fashion/exes-explain-ghosting-the-ultimate-silent-treatment.html?_r=0. Accessed 1 March 2020.

Sartre, Jean-Paul (1966), *Being and Nothingness*, New York: Washington Square Press.

Schull, Natasha D. (2014), *Addiction by Design: Machine Gambling in Las Vegas*, Princeton: Princeton University Press.

Sharma, Sarah (2014), *In the Meantime: Temporality and Cultural Politics*, Durham: Duke University Press.

Steiner, Henriette and Veel, Kristin (2020), *Tower to Tower: Gigantism in Architecture and Digital Culture*, Cambridge: MIT Press.

'tick', *The Oxford English Dictionary*, Oxford: Oxford University Press.

Thylstrup, Nanna Bonde and Teilmann, Stina (2017), 'Thumbnail images: Uncertainties, infrastructures and search engines', *Digital Creativity*, 28:4, pp. 279–96.

Tiffany, Kaitlyn (2019), 'The Tinder algorithm explained: Some math-based advice for those still swiping', *Vox*, 18 March, https://www.vox.com/2019/2/7/18210998/tinder-algorithm-swiping-tips-dating-app-science. Accessed 1 December.

Veel, Kristin (forthcoming 2021), 'Social media small forms and flirtatious storytelling practices', in B. Brandl-Risi and L.

Ruprecht (eds), *Literatur & Performance*, Berlin: de Gruyter.

Veel, Kristin and Thylstrup, Nanna Bonde (2018), 'Geolocating the stranger: The mapping of uncertainty as a configuration of matching and warranting techniques in dating apps', *Journal of Aesthetics and Culture*, 10:3, pp. 43–52.

Vogl, Joseph and Ritter, Nils. N.d. 'The literary and epistemic history of small forms', *Kleine Formen*, https://www.doctoral-programs.de/browse-programs/social-sciences-and-humanities/the-literary-and-epistemic-history-of-small-forms/. Accessed 1 March 2020.

Wohn, Yvette Donghee, Carr, Caleb T. and Hayes, Rebecca A. (2016), 'How affective is a "like"? The effect of paralinguistic digital affordances on perceived social support', *Cyberpsychology, Behavior, and Social Networking*, 19:9, pp. 562–66.

Zylinska, Joanna (2017), *Nonhuman Photography*, Cambridge: MIT Press.

Note

1. Leap seconds are a good example of just how vulnerable to uncertainty time measurement systems are. Leap seconds are inserted by the International Earth Rotation and Reference Systems Service to make up for the difference between the precise time measured by atomic clocks and observed solar time. This difference is due to the fact that the Earth's rotation speed can vary in response to climatic and geological events. The last time a leap second was added was in 2016. However, the irregularity and unpredictability of leap seconds causes significant challenges, glitches and outages for computing systems, and becomes particularly critical in high-speed trading and automated systems.

Contributors

Adam Basanta is an artist, composer and performer of experimental music. Born in Tel-Aviv (ISR) and raised in Vancouver (CAN), he lives and works in Montreal. In his installation works, Basanta arranges common commercially available objects into delicately intertwined and seemingly performative choreographies, disrupting their technical and economic functions while revealing their material agencies and status as extended technological prostheses. His work has been recently exhibited in galleries and institutions including Carroll/Fletcher Gallery (UK), Galerie Charlot (FR-ISR), Arsenal Contemporain (CAN), Fotomuseum Winterthur (CH), National Art Centre Tokyo (JPN), American Medium Gallery (NYC), New Media Gallery (CAN), V Moscow Biennale for Young Art (RUS), Serralves Museum (POR), Edith-Russ-Haus fur Mediakunst (GER), Villa Brandolini (ITA), Vitra Design Museum (GER), York Art Gallery (UK) and The Center for Contemporary Arts Santa Fe (USA).

Andrew Blanton is an assistant professor and area coordinator of the CADRE Media Labs at San Jose State University and a Ph.D. student in music composition working at the Center for New Music and Audio Technologies (CNMAT) at the University of California Berkeley. His work has been performed and presented around the world in venues such as Google Cultural Lab in Paris, the University of Brasilia, the City University of Hong Kong and STEIM Amsterdam, among many others. His current work focuses on the emergent potential between cross-disciplinary arts and technology in the context of Composition, New Media Art and building sound + visual environments through software development. Andrew has advanced expertise in percussion, 3D environments/graphics programming, creative software development and developing projects in the confluence of art and science.

Lynn Comella is an associate professor of gender and sexuality studies at the University of Nevada, Las Vegas. As an expert on the adult entertainment industry, she has written extensively about sex and culture for academic publications and popular media outlets. She is the author of *Vibrator Nation: How Feminist Sex-Toy Stores Changed the Business of Pleasure* (2017), which was a *New York Times* Editors' Choice, and a co-editor of *New Views on Pornography: Sexuality, Politics and the Law* (2015). Her research has been featured in the *New York Times*, *Washington Post*, *Rolling Stone*, *The Atlantic* and more.

Olga Fedorova is a Russian media artist with an MA in painting from ENSAV, Brussels, working at the intersection of photography, painting and digital imaging. She has exhibited in solo and group shows all over Europe. Her work has been recently presented at Annka Kultys Gallery (London), Kunsthaus Langenthal (Bern) and Galerie Charlot (Paris). She has also been included in group shows at renowned art galleries such as In De Ruimte (Ghent), Pulsar (Antwerp) and Russiantearoom Gallery (Paris). In 2016, Fedorova released a solo show « The Inevitability of a Strange World » at Liebaert Projects in Kortrijk, Belgium, as well as a virtual solo exhibition at offspace.xyz. Her video work has been part of virtual exhibitions for The Wrong Biennale, dadaclub.online, Felt Zine and Blockedart.com. She currently lives and works in Brussels, Belgium.

Zach Gage is a New York game designer, programmer, educator and conceptual artist. His work often explores the powerful intersection of systems and social dynamics, which he rehearses by means of interrogating existing systems in digital spaces and by offering entirely new game-based systems. An Eyebeam Alumni, Apple Design and Game of The Year Award Winner and BAFTA Nominee, he has exhibited internationally at the Venice Biennale, the New York MoMA, The Japanese American National Museum in Los Angeles, XOXO Festival in Portland, FutureEverything in Manchester, The Centre for Contemporary Art Ujazdowski Castle in Warsaw and in Apple stores worldwide. His work has been featured in a number of online and printed publications, including *The New York Times*, *Art in America*, *The New York Times Magazine*, *EDGE Magazine*, Rhizome.org, *Neural Magazine*, *New York Magazine* and *Das Spiel und seine Grenzen* (Springer Press).

Tom Galle is a Belgian conceptual artist working in the realm of the surreal and the internet-inspired environments. His fast-paced conceptual art keeps up with the speed of the internet to reflect (and also to set up) the accelerated modes of existence. His viral meme cycles are often a catalyst for discussions on subjects of digital zeitgeist. Tom's visual language is best known for its simplicity, absurdism and sarcasm. Its recurring themes include meme-culture phenomena, digital intimacy, internet dependency, contemporary/corporate branding and other. After living few years in New York, Tom currently lives and works between Brussels and Berlin.

Pablo Garcia is an associate professor in the Department of Contemporary Practices at the School of the Art Institute of Chicago. Trained as an architect, Pablo has moved from design-for-hire to internationally exhibited artist, whose work embraces provocations and research. Previously, Pablo has taught at Carnegie Mellon University, Parsons School of Design and the University of Michigan. From 2004 to 2007, he also worked as an architect and designer for Diller Scofidio + Renfro. He holds architecture degrees from Cornell and Princeton Universities.

Thomas Israël is a Brussels-based multimedia artist preoccupied with immersive, interactive video installations, sculptures, video scenography for stage and non-stage performances. Having begun his career in theatre, he evolved towards digital arts offering an atypical approach that revolves around the themes of the body, time and the subconscious. His work has been shown at MoMA in New York, the Society for Arts and Technology in Montreal, the Musée des Abattoirs in Toulouse and at many festivals, exhibitions, galleries and museums across the world. His monograph *Memento Body* was launched at la Lettre Volée in 2015, and his performances in body-mapping (that made him a laureate of the prestigious Japan Media Art Festival in 2014) are touring worldwide. His work is represented by Galerie Charlot.

Karen Lancel and Hermen Maat are artists and researchers who pioneer in exploring embodied presence, intimacy, privacy and trust in the posthuman bio(techno)logical entanglements with (non-)human others. They radically re-orchestrate automated control technologies, neuro-feedback and disrupted sensory perception to create the so-called Trust-Systems. Their immersive audiovisual performances *Meeting Rituals* and *Reflexive Datascapes* are based on social multi-sensory experience of AI, brain and touch communication, for shared emotion, reflection and public dialogue in merging realities with ecologies. Their performance-installations and video works have been shown internationally: 56th Venice Biennale 2015, ZKM Karlsruhe, Ars Electronica, Transmediale Berlin, Stedelijk Museum Amsterdam, ISEA 2004–19, HeK Basel, RIXC Riga, TASIE Beijing 2006–19, National Museum of China; supported by Mondriaan Fund, NWO Netherlands Research Fund, V2Lab Rotterdam, European Media-Art Platform/Creative Europe Program (EMAP/EMARE). Lancel is a Ph.D. researcher at TU Delft, previously heading MFA media-art Hanze at the University

Groningen where Maat is a Ph.D. researcher and lecturer. They have published internationally with *CHI 2018*, Springer, *Leonardo*, MIT. The have been granted research fellowships at Banff Center Canada, Vienna University, TASML Beijing. Their works are included in public and private collections (e.g. ZKM Karlsruhe and LIMA Foundation, Amsterdam). They have won a number of awards, including GAAC Global AI Art Competition 2019 Beijing, Virtueel Platform 2010–12, Film Fund Golden Calf Interactive nomination 2019, TASIE2019 Wu Guanzhong Award/Tsinghua University Beijing.

Jeroen van Loon is a Utrecht-based artist with a bachelor degree in Digital Media Design and a master's degree from the HKU University of the Arts Utrecht. van Loon's work has been shown in solo and group exhibitions that have earned him a European Youth Award and a KF Hein Art Grant. He regularly gives talks on his artistic explorations of technology for the art world and through institutions that promote innovation, such as TEDx. His recent work has become part of the Verbeke Foundation Collection in Belgium. He has exhibited at 'Alien Matter' (transmediale, HKW, Berlin), 'Beyond Data' (Central Museum, Utrecht), Dutch Design Week (Eindhoven), 'Design my Privacy' (Z33 Hasselt, Belgium), Cyberfest 9 hosted in Russia/USA/Colombia and V2Lab in Rotterdam.

Kyle Machulis better known as qDot, is a graduate of the University of Oklahoma with a degree in computer science. He currently spends his days as a mild-mannered engineer. Kyle has worked in everything from sex and technology to health devices to self-driving cars to device reverse engineering and more (see the portfolio page at https://kyle.machul.is/portfolio/).

Lee Mackinnon is a senior lecturer at the University of the Arts London and is currently MA Photography course leader. Lee's research is largely situated in fields of comparative media studies and visual technology. Her writing explores how technical systems affect the way we see, think and communicate. Central to this work is consideration of the ways human bias and inequality are perpetuated by technical systems, despite claims to the contrary. Lee's writing has appeared in *e-flux*, *Third Text* (Routledge) and *Leonardo* (MIT Press). Her artwork has featured in exhibitions at The Bloomberg Space and Camden Arts Centre. Recent commissions include artwork for the Japanese pop group, *Viva Sherry* (mottomotto records).

Ania Malinowska is an associate professor at the University of Silesia (Poland) and a former Senior Fulbright Fellow at the New School of

Social Research in New York. She is a coeditor of (with Karolina Lebek) *Materiality and Popular Culture: The Popular Life of Things* (Routledge, 2017), (with Michael Gratzke) *The Materiality of Love: Essays of Affection and Cultural Practice* (Routledge, 2018) and (with Toby Miller) 'Media and Emotions. The New Frontiers of Affect in Digital Culture' (a special issue of *Open Cultural Studies*, 2017). She has authored many papers and chapters in cultural and media studies with regard to love, social norms, codes of feelings and technology. She is currently working on a mono-graph *Love in Contemporary Technoculture* (under contract with CUP).

Andrew McStay is a professor of digital life at Bangor University, UK. His recent book, *Emotional AI: The Rise of Empathic Media* examines the impact of technologies that make use of data about affective and emotional life. Current projects include the study of emotional AI, chil-dren and parents and (separately) cross-cultural analysis of emotional AI in the United Kingdom and Japan. Non-academic work includes IEEE membership (P7000/7014) and ongoing advising roles for start-ups, NGOs and policy bodies. He has also appeared and made submissions to the United Nations Office of the High Commissioner on the right to privacy in the digital age, the UK House of Lords AI Inquiry and the UK Department for Culture, Media and Sport Inquiry on fake news and reality media.

Derek Conrad Murray is an interdisciplinary theorist specializing in the history, theory and criticism of contemporary art and visual culture. He works in contemporary aesthetic and cultural theory with a particular attention to technocultural engagements with identity and representation. He is currently a professor of History of Art and Visual Culture at the University of California, Santa Cruz. Murray serves as an associate editor of *Nka: Journal of Contemporary Afri-can Art*. He is also currently serving on the Editorial Board of *Art Journal* (CAA) and the Editorial Advisory Board of *Third Text*. Murray is the author of *Queering Post-Black Art: Artists Transforming Afri-can-American Identity after Civil Rights* (I.B. Tauris, UK, 2016) and is also the author of two forthcoming volumes: *Mapplethorpe and the Flower: Radical Sexuality and the Limits of Control* (Bloomsbury, 2020) and an edited volume entitled *Visual Culture Approaches to the Selfie* (Routledge Press, 2020).

David Parisi is an associate professor of emerging media in the Department of Communication at the College of Charleston. His book *Archaeologies of Touch: Interfacing with Haptics from Electric-ity to Computing* (University of Minnesota Press, 2018) investigates the past, present and possible futures of technologized touch, weaving together accounts of tactility from psychophysics, cybernetics, electrotherapy, virtual reality, cybersex and mobile communication to provide a comprehensive overview of the ways that touch has been radically transformed by its encounters with technology and science. He is also a coeditor of the *New Media & Society* special issue on Haptic Media Studies, and his research on touch has been featured in *ROMchip: A Journal of Game Histories, Convergence, Game Studies, The Wall Street Journal, Vice, Playboy, Logic Magazine, Immerse* and the podcasts *Stroke of Genius, All in the Brain* and *INT: A Podcast on the Tactile Internet*.

Valentina Peri is an art curator and cultural anthropologist with expertise in new media and digital art. For ten years she has run Galerie Charlot with locations in Paris and Tel Aviv, where she has exhibited projects focused on new media art. She is also a co-founder of SALOON Paris, an international network of woman identifying art professionals. Valentina's interests focus on the impact of technology on contemporary culture. Exhibitions she has curated include Ben Grosser, « Systems under Liberty », 2015; « Archeonauts », a group show featuring works of artists and activists as Morehshin Allahyari, Eduardo Kac, Disnovation.org, Evan Roth among others, who shares an « archaeological gaze » on the present, in 2017; Quayola « Remains », 2018; Nicolas Sassoon « Subterranea », 2020. She is an initiator and curator of « Data Dating », a group show about love and desire in the internet age that features the work of renowned international media artists and which has been shown in Paris (2018), Tel Aviv (2019) and London (2020).

Lauren Rosewarne is a senior lecturer in the School of Social and Political Sciences at the University of Melbourne, Australia. She regularly comments and speaks on a wide variety of topics including gender, sexuality, politics, public policy, social media, pop culture and technology. Lauren is the author of eleven books including *Why We Remake: The Politics, Economics and Emotions of Film and TV Remakes* (2020); *Sex and Sexuality in Modern Screen Remakes* (2019); *Analyzing Christmas in Film: Santa to the Supernatural* (2018); *Intimacy on the Internet: Media Representations of Online Connections* (2016); *Cyberbullies, Cyberactivists, Cyberpredators: Film, TV, and Internet Stereotypes* (2016); *Masturbation in Pop Culture: Screen, Society, Self* (2014); *American Taboo: The Forbidden Words, Unspoken Rules, and Secret Morality of Popular Culture* (2013); *Periods In Pop Culture: Menstruation in Film and Television* (2012); *Part-Time Perverts: Sex, Pop Culture and*

Kink Management (2011); *Cheating on the Sisterhood: Infidelity and Feminism* (2009) and *Sex in Public: Women, Outdoor Advertising and Public Policy* (2007).

Gilad Rosner is a privacy and information policy researcher and founder of the non-profit IoT Privacy Forum. Gilad's work is focused on identity management, privacy governance and emerging technologies. Gilad is a member of the UK Cabinet Office Privacy and Consumer Advisory Group and the Advisory Group of Experts convened to support the forthcoming review of the OECD Privacy Guidelines. He is a visiting researcher at the Horizon Digital Economy Research Institute and the co-founder of Eleos, a consultancy providing ethics and privacy services to companies deploying emotion analytics.

Moises Sanabria is a media artist born in Caracas, Venezuela, with interest in technology, internet and contemporary branding. He holds a BFA from New York's Cooper Union School of the Arts. He is best known for his work around online culture, most notably the work done under the collective ART404 '5 Million Dollars 1 Terabyte' – a sculpture consisting solely of a 1 TB Black External Hard Drive, containing just under $5,000,000 worth of illegally downloaded files. Moises is concerned with the language of online and meme culture, especially with the sense of familiarity and distance they create. His favourite art forms are online stunts, websites, apps, performances or physical sculptures and hardware installations. Moises's work is often re-purposed and re-contextualized by press and numerous internet accounts. He currently lives and works between Brooklyn and Miami.

Antoine Schmitt is an installation artist whose work addresses the processes of movement and question their intrinsic conceptual problematics, of plastic, philosophical or social nature. Heir of kinetic and cybernetic art, nourished by metaphysical science fiction, his work interrogates the dynamic interactions between human nature and the nature of reality. Trained as a programming engineer in human computer relations and artificial intelligence, he uses programs that he develops himself as a contemporary artistic material and places them at the core of most of his artworks. Schmitt has received a number of awards from international new media art festivals such as transmediale, Ars Electronica, Vida 5.0. He has been exhibited at the Centre Georges Pompidou, MAD Paris, Sonar Barcelona, Ars Electronica, CAC of Siena, MAC Lyon. His work is part of notable private and public art collections such as FMAC (Paris), Artphilein Foundation (Vaduz, Switzerland), Frankel

Foundation For Art (Bloomfield Hills, USA), Société générale (Paris). Antoine lives and works in Paris and is represented by Galerie Charlot.

Nanna Bonde Thylstrup is an associate professor of communication and digital media at Copenhagen Business School. Her research interests focus on knowledge and ignorance infrastructures, environmental media and digital epistemologies. Her works specifically concentrate on how media theory, cultural theory and critical theory unpack and unfold problems of datafication and digitization with regard to race, gender and sexuality. She has authored *The Politics of Mass Digitization* (MIT, 2019) and co-edited *Uncertain Archives. Critical Keywords for Big Data* (MIT, 2020). She has written widely on the politics of infrastructures, the gendered epistemologies of veracity, the effects of self-tracking and the ethics of data waste as well as contributed to exhibitions and artistic research practices.

Kristin Veel is an associate professor at the Department of Arts and Cultural Studies, University of Copenhagen. She holds a Ph.D. in German literature from the University of Cambridge, UK. Her research focuses on the impact of information and communication technology on the contemporary cultural imagination, with a particular interest in issues of information overload, surveillance, invisibility and uncertainty and the way in which these are negotiated in film, art and literature. She is a co-author of *Tower to Tower: Gigantism in Architecture and Digital Culture* (MIT Press, 2020) and has previously published the monograph *Narrative Negotiations: Information Structures in Literary Fiction* (Vandenhoeck & Ruprecht, 2009). She is a co-editor of ten collected volumes and special journal issues. Together with Nanna Bonde Thylstrup, she has published a series of articles on dating apps as cultural techniques.

Addie Wagenknecht is a new media artist whose work explores the tension between expression and technology by seeking to blend conceptual work with forms of hacking and sculpture. Wagenknecht's work has been presented at MuseumsQuartier Wien (Vienna), La Gaîté Lyrique (Paris), The Istanbul Modern (Istanbul), Whitechapel Gallery (London) and MU (Eindhoven). In 2016, she collaborated with Chanel and *I-D magazine* as part of their Sixth Sense series. In 2017, her work was acquired by the Whitney Museum for American Art in New York. Her work has been featured in a number of books and magazines such as *TIME*, *Wall Street Journal*, *Vanity Fair*, *Art in America* and *The New York Times*. Wagenknecht holds a Master's degree from the Interactive Telecommunications Program at New York University and has previously held fellowships at Eyebeam Art + Technology Center in New York City, Culture Lab in Newcastle-upon-Tyne, UK, Institute HyperWerk for Postindustrial Design in Basel and The Frank-Ratchye STUDIO for Creative Inquiry at Carnegie Mellon University. She is represented by bitforms gallery in New York City.

John Yuyi born in Taiwan in 1991, John Yuyi graduated from Shih Chien University in Taipei with a degree in Fashion Design and currently lives and works in New York. She belongs to the millennial generation, and her work is deeply rooted in and inspired by the internet culture. Behind the camera, she is a photographer, fashion designer and art director. In front of the camera, she is a model and performance artist. Her work explores everyday internet symbols and phenomena in relation to the young bodies of the internet users to comment upon the hidden yet candid anxiety of her generation, marked by the worship of new media and the search for identity. Her work owes to social media platforms like Instagram, Facebook and Weibo whose ecosystems and templates trained her to become sensitive about visual expression as well as social, market and design trends. John Yuyi captured a crowd in a very short amount of time and became a new favourite of brands such as Gucci and NIKE, with whom she has launched several projects and commercial campaigns. Being a photographer, a model, a designer and an artist, John Yuyi gradually separates herself from an artist's own identity through constant superposition and conversion, expanding the possibilities of constructing and reorganizing her identity.

!Mediengruppe Bitnik (read – the not Mediengruppe Bitnik) is a Berlin-based conglomerate of artists working on, and with, the internet. Their practice expands from the digital to physical spaces, often intentionally applying loss of control to challenge established structures and mechanisms. Recent works by the artists have involved subverting online surveillance cameras or bugging an opera house to illegally broadcast its concerts outside. In 2013, they sent a parcel with a camera to the WikiLeaks founder Julian Assange at the Ecuadorian embassy in London to broadcast the journey through the postal system live on the internet. The group sparked controversy for creating a bot called *Random Darknet Shopper* and sending it on a three-month shopping spree in the dark-net where it randomly bought items like cigarettes, keys and Ecstasy, before having them sent directly to a gallery in Switzerland. Their work is shown internationally, most recently in exhibitions at Annka Kultys Gallery in London, House of Electronic Arts in Basel, Galerie Charlot in Paris, Eigen + Art Lab in Berlin, Super Dakota in Brussels, Centre Culturel Suisse in Paris, Aksioma in Ljubljana,

Kunsthaus in Zurich, FACT in Liverpool, Onassis Cultural Center in Athens, Public Access Gallery in Chicago, Kunstverein in Hannover, Nam June Paik Art Center in South Korea, Fondazione Prada in Milano, Shanghai Minsheng 21st Century Museum, The Pushkin Museum of Fine Arts in Moscow, Cabaret Voltaire in Zurich, Beijing Contemporary Art Biennial and the Tehran Roaming Biennial.

Index